Supply Chain in Agri-Business

Supply Chain in Agri-Business

Authors:

Anil Sharma

Naveed Hamid

Jyoti Kachroo

SATISH SERIAL PUBLISHING HOUSE

403, Express Tower, Commercial Complex

Azadpur, Delhi-110033 (India)

Phone : 011-27672852, Fax : 91-11-27672046

e-mail : info@satishserial.com, hkjain1975@yahoo.com

Website : www.satishserial.com

Published by :

SATISH SERIAL PUBLISHING HOUSE
403, Express Tower, Commercial Complex, Azadpur, Delhi-110033 (INDIA)
Phone : 011-27672852 Fax : 91-11-27672046
E-mail : info@satishserial.com, hkjain1975@yahoo.com

ISBN 978-93-88020-07-7

Composed, Designed & Printed in India

About the Authors

Anil Sharma possesses a sound and brilliant academic record in Masters (Agri-Business Management) Programme. He has obtained his degree in Agricultural Business Management from the Sher-e-Kashmir University of Agricultural Sciences and Technology of Jammu. He remained good in his master's degree programme. He has 2 research publications to his credit.

Naveed Hamid possesses a meritorious academic record and awarded by university merit certificate in recognition of his having achieved the prescribed scholastic standards of high order OGPA in Masters (Agri-Business Management) Programme. He has obtained his degree in Agricultural Business Management from the Sher-e-Kashmir University of Agricultural Sciences and Technology of Jammu. He has 26 research publications in referred and UGC journals, 10 abstracts, 02 books and 4 international conferences to his credit. He is a member and reviewer of various agri-business, economics & management journals. He is awardee of a prestigious project by NABARD in 2014 and three awards from repute organization for his contribution to the multidisciplinary field. He earned Young Scientist award by ITSR foundation for his remarkable achievements in Agri-Business Field. His field of expertise is Agricultural Business Management, Supply chain/Value chain and marketing. Naveed Hamid is presently working as Assistant Professor©, Agri-Business Management in Division of Agricultural Economics & ABM, SKUAST Jammu. In his two and a half years experience in teaching and research, he has guided six research PG Scholars.

Dr. Jyoti Kachroo has remained gold medalist (Two) and meritorious throughout her academic career, graduated and post-graduate (Economics) from University of Jammu, Jammu and Kashmir. She was initially appointed as lecturer (Economics) in Govt. Degree Collage, Jammu and later joined the division of agricultural Economics and Statistics, SKUAST, Jammu as Associate Professor. Presently she is Professor and Head, Division of Agricultural Economics & Agri-Business Management. Her field of expertise is women empowerment, agricultural production economics, marketing and econometric model based research. She has published about 85 articles which include chapters, popular articles, review papers and about 40 research publications in referred journals. She has been honoured with D.T.Doshi Foundation award for best presentation. She has 24 years experience of teaching and research and she has guided 16 research scholars.

Preface

Supply Chain Management (SCM) is an essential element to get operational efficiency. Supply Chain Management can be applied to customer satisfaction and company success, as well as within societal settings, including medical missions, disaster relief operations and other kinds of emergencies, cultural revolution, and it can help improve quality of life. Because of the vital role Supply Chain Management plays within organizations, employees seek employees with an abundance of Supply Chain Management skills and knowledge. Supply chain management is helpful in reducing costs. It provides better medium for information sharing between partners. It is also helpful in improving customer's satisfactions as well as services. Supply chain management is also helpful in increasing cash flow and improves quality and gives higher profit margin.

The various chapters of the book focus on the supply chain process and its proper management. The book also throws light on the aspect of supply chain and objectives on which supply chain works within the organization.The book mainly focuses on the issues related to the supply chain regarding the basmati rice brand mill. This also explains the different supply chain models currently existed for the supply of basmati rice within in the particular organization. The next chapter throws light on the research methodology and tools adopted for analysis to get the desired result related to supply chain and its main constraints and problems.

We felt our heartfelt gratitude to Dr. Kachroo (Registrar SKUAST-J), whose genuine guidance and conceptual technical ideas during the course of planning and timely help in finalizing this book.

List of Tables

List of Figures

List of Flowcharts

Contents

Chapter **1**

Importance and Scenario of Basmati Rice

Milling is the process wherein the rice grain is transformed into a form suitable for human consumption, therefore, has to be done with utmost care to prevent breakage of the kernel and improve the recovery. Brown rice is milled further to create more visually appealing white rice. After harvesting and drying, paddy is subjected to the primary milling operation which includes de-husking as well as the removal of bran layers (polishing), before it is consumed. In this process, the rice which is obtained after milling is called raw rice. Another process through which rice is obtained after milling is called "Parboiling Rice." Nearly 60 per cent of the total rice produced in India is subjected to parboiling. Rice milling losses may be qualitative or quantitative in nature. Quantitative or physical losses are manifested by low milling recovery while low head rice recovery or high percentage of broken kernel reflects the qualitative loss in rice grains.

Rice has potential in a wide range of food categories. Besides having nutritional and medicinal benefits, the by-products of rice are equally important and beneficial. By-products from growing rice create many valuable and worthwhile products. The inedible parts which are discarded through the milling process and the edible part could be transformed into some of the following suggested products like rice husks, rice bran, broken rice, rice flour, rice milk, rice pudding, rice starch, rice straw. Rice is also used in beverage making, rice paper, rice glue, rice cakes, rice vinegar, rice soya milk and red yeast. Apart from every household, there are many bulk buyers like hospitals, caterers, restaurants, hostels, etc. With proper efforts, it is possible to enter into a long term contract with big traders or contractors.

India is the world's second largest producer of Rice, Wheat and other cereals. The huge demand for cereals in the global market is creating an excellent environment for the export of Indian cereal products. In 2008, India had imposed ban on export of rice and wheat, etc. to meet domestic needs. Now, seeing the huge demand in the global market and country's surplus production, Country has lifted the ban, but only limited amount of export of the commodity are allowed. The allowed marginal quantity of exports cereals could not make any significant impact either on domestic prices or the storage conditions. The important cereals are - wheat, paddy, sorghum, millet (Bajra), barley and maize etc. According to the final estimate for the year 2011-12 by ministry of agriculture of India, the production of major cereals like rice, maize and Bajra stood at 105 million tonnes, 21.76 million tonnes and 10.28 million tonnes respectively. (APEDA)

India is not only the largest producer of cereal as well as largest exporter of cereal products in the world. India's export of cereals stood at ₹ 58279.80 crores during the year 2014-15. Rice (including Basmati and Non-Basmati) occupy the major share in India's total cereals export with 64.40 percent during the same period. Whereas, other cereals including wheat represent 35.60 percent share in total cereals exported from India during this period. The major importing countries of India's cereals during the period were Iran, Saudi Arabia, Indonesia, United Arab Emirates and Bangladesh.

1.1 Varieties of Basmati Rice

The main varieties of Basmati rice as notified under the seeds Act, 1966 are Basmati 386, Basmati 217, Ranbir Basmati, Karnal Local/ Taraori Basmati, Basmati 370, Type-3 (Dehradooni Basmati), Pusa Basmati-1, Pusa Basmati 1121, Punjab Basmati-1, Haryana Basmati-1, Kasturi and Mahi Sugandha.

1.2 Rice Exports from India

The steady increase in production, availability of buffer stocks and the growing demand for basmati rice in the international market made India an important rice exporting country of the world. In the year 2008-09 India exported 2.48 Metric Tonnes of rice earning very high foreign exchange to the tune of ₹ 11164 crores. Of the several agro products exported from India, rice alone constituted 30.75 per cent of foreign exchange in the year 2008-09 and India is the fourth major exporter of rice after Thailand, Vietnam and USA. For a long time India was not a rice exporting country. The major production and productivity gains have enabled the country to attain self-sufficiency in rice since mid 1980's. Indian exporters entered into the world rice trade largely through the export of small quantities of highly priced basmati rice which was less than 5 per cent in mid 1980's. In fact, the global rice trade at that time was also low, only 4 per cent which was 11.9 Metric Tonnes of the world rice production of 265.9 Metric Tonnes (on milled rice basis). The world rice trade quantum also doubled by 2001 to 27.9 Metric Tonnes and by 2008 to 29.0 Metric Tonnes, of which India's share is around 8.62 per cent, while in 2006-07, it was the highest (19.8 per cent).

The quantum of total rice exports from India was mere 0.38 Metric Tonnes in 1987-88, which rose to 0.53 Metric Tonnes in 1990-91 and increased phenomenally to a record scale of 5.51 Metric Tonnes in 1995-96. It dropped to around 2 Metric Tonnes during 2001 and again rose to as high as 6.46 Metric Tonnes in 2007-08, which is a twelve times increase over the quantum India exported in 1990-91. Similarly, the value too rose from ₹ 456 crore in 1992 to ₹ 11164 crore in 2008-09 which is a spectacular twenty four times increase in foreign exchange earnings.

During the last 5 years, India was within the first five rice exporting countries with Thailand consistently ranking first in world rice exports with 10.00 Metric Tonnes in 2008. The other major exporters include Vietnam (4.65 MT), USA (3.50 MT), India (3.30 MT), Pakistan (3.0 MT), China (0.95 MT), Uruguay (0.78 MT) and others (4.02 MT) (USDA 2008). Major Rice importing countries are Nigeria, Philippines, Iran, Saudi Arabia, Iraq, EU, South Africa, Malaysia, Bangladesh, Indonesia and The Ivory Coast.

India is the leading exporter of the basmati rice to the global market. The country has exported 37,02,2284 Metric Tonnes of basmati rice to the world for the worth of ₹ 27,597.87 crore during the year 2014-15. India has been top exporter in global rice trade. The sustainability in rice exports is the only one with a competitive-edge internationally as compared to other Indian agro commodities is the resultant outcome of a combination of external factors, dynamics of domestic market, hybridisation of paddy, and efficient execution of contracted business both from east and west coast ports of India.

Between August and September 2014, according to rice exporters, about 80 per cent of the exports were shipped to Iran. Iran's annual demand for basmati rice is close to 2.4 million tonnes (MT), and its own production is 1.4 MT. This means, an annual import requirement of 1 MT. Thus, Iran needs to import about 100,000 Metric Tonnes of basmati rice every month to meet its demand. Traditionally, basmati rice has been among India's top Agri-export commodity and Iran, the biggest buyer. Following Iran's reluctance to buy basmati rice from India, exports have taken a major hit. In 2014-15, India exported 0.93 MT of basmati rice to Iran compared with 1.4 MT in 2013-14, according to Agricultural and Processed Food Products Export Development Authority. The realisation from basmati rice exports to Iran in 2014-15 was ₹ 6,759 crore, against ₹ 10,976 crore in 2013-14, a fall of 60 per cent. The duty on import of rice has been increased by Govt. of Iran from 22 per cent to 40 per cent in view of the domestic crop season. The hike will continue till January 21, 2015. The hike in customs duty during the crop season is a regular feature in Iran. Moreover, Iran has imposed a temporary restriction on import of rice with effect from 19th October, 2014. The Government engages with its trade partners in an institutional basis and takes up all trade related issues from time to time.

Even though, the volume of basmati rice exports increased by 16.92 per cent during April-August 2015 fiscal year, its exports in value terms declined by 23.15 per cent over the same period last year. Despite a surge in export volume, the overall realization remained less by a significant margin. The drop in realization was attributed to the lower average price of basmati rice, which declined to US$ 929 per tonne during April-August 2015 from US$ 1,415 per ton a year ago. The slowdown was triggered by several domestic as well as international factors, including weak international demand, poor quality and huge carryover stock. India has exported 55.26 Lakh tonnes of rice, valued at over US$ 3.17 billion, in the first six months of 2015. This shows that with the upcoming year, basmati is increasing its popularity in Indian as well as the overseas market, which is increasing the production as well as the price of the product.

1.3 Basmati brief Scenario

Rice is one of the most crucial food crops in the world and a staple diet for nearly half the global population. Over 90 per cent of the global rice output and consumption is centered in Asia, wherein the world's largest rice producers, China and India, are also the world's largest rice consumers. India accounts for over 70 per cent of the world's basmati rice production. Basmati rice constitutes a small portion of the total rice produced in India. By volume, the share of basmati rice is around 6 per cent in 2014-15, even as by value, basmati rice exports account for 57 per cent in 2014-15, of India's total rice exports. Basmati rice exports have increased at a compounded annual growth rate (CAGR) of 27 per cent from ₹ 28.24 billion in 2004-05 to ₹ 275.98 billion in 2014-15. The proportion of basmati rice exports in India's total exports has increased from around 0.6 per cent to around 1.3 per cent during the last one decade. While basmati rice is consumed across the globe, West Asian countries account for 75 per cent of Indian basmati rice exports in 2014-15. Within West Asia, Iran and Saudi Arabia are the two largest buyers, together accounting for over 50 per cent of basmati rice exports from India.

However, even as Iran emerged as one of the largest importers of basmati rice in recent years, the country imposed a ban on basmati rice imports from India in 2014-15, citing its own healthy rice crop and large basmati inventory. Commenting on pricing scenario, ICRA said basmati paddy is also vulnerable to cyclical price fluctuations. Higher prices in the market encourage higher basmati paddy cultivation, which increases supply in the next season. This depresses the price, thereby erasing gains and shifting farmers away from basmati paddy cultivation. During the procurement season of 2012-13 and 2013-14, there was a steep rise in paddy prices from around ₹ 18,000 per tonne (Metric Tonnes) in 2011-12 to around ₹ 37,000 Metric Tonnes in 2013-14, due to strong demand in the international market. (Agri.net) and (APEDA)

1.4 Indian states Contribution in Basmati Rice Production and Export

Basmati cultivation is confined to traditional basmati growing areas in North-west Indian states - Punjab, Haryana, Uttarakhand and Western Uttar Pradesh and to a limited area in Himachal Pradesh, Delhi, and Jammu & Kashmir. Localized cultivation of basmati rice is encountered in Karnal, Panipat, Kurukshetra, Jind, Kaithal, Ambala, Yamuna Nagar and Sonipat districts in Haryana, Amritsar, Gurdaspur, Kapurthala, Jalandhar, Patiala, Ropar, Nawan Shehar, Fatehgarh Sahib and Hoshiarpur in Punjab; Dehradun, Nainital, Udhamsingh Nagar, Haridwar in Uttarakhand; Pilibhit, Saharanpur, Rampur, Bijnor, Muzzaffar Nagar, Moradabad and Badaun in Uttar Pradesh; Kullu, Kangra, Solan, Mandi and Sirmaur in Himachal Pradesh; Jammu and Kathua in Jammu & Kashmir. Of the 25 to 29 Metric Tonnes of rice traded annually in the world market, basmati rice market is less than 10 per cent. But basmati rice captures higher returns, as it is priced three times higher (US$ 800-1200 per metric tonne) over non-basmati rice (US$ 200-400 per metric tonne) in the international as well as in the domestic markets. About 50-70 per cent of basmati produced in the country is exported mainly to Saudi Arabia (68 per cent), United Arab Emirates, United Kingdom, Kuwait, Bahrain and others. With a small beginning in 1978-79, India exported 67,000 tonnes of basmati rice earning a modest foreign exchange of ₹ 32 crore. The upward trend continued steadily with the quantum of basmati rice raising from 470,000 tonnes in 1996-1997 to 770,764 tonnes in 2003-2004 and the value from ₹ 1200 crore to a maximum of ₹ 1991 crore during the same period. Highest quantum of 1.55 MT of basmati rice was exported in 2008-09 earning maximum foreign exchange to the tune of ₹ 9477 crore.

Non-basmati rice has also become a thrust area for export promotion, though there have been fluctuating trends. It is heavily dependent on the policy changes of the government from time to time and has experienced quota regimes to liberalization. From a mere 28,422 tonnes during triennium ending 1987-89, non-basmati exports rose to a record level of 0.53 MT during 1991-92 but declined again to 0.32 MT during the triennium 1992-94. The highest quantum exported was 5.12 Metric Tonnes in 1995-96 earning foreign exchange amounting to ₹ 3702 crore. Although during the year 2007-08, India exported 5.285 Metric Tonnes of non-basmati rice earning ₹ 7409 crore the quantum and value dropped in 2008 due to the ban imposed by the Government of India on export of non-basmati rice to build up the domestic buffer stocks. (APEDA)

In 2014-15, the total area under Basmati cultivation was 2.1 million hectares, compared to 1.6 million hectares in 2013-14, an increase of 31 per cent. Basmati Rice production contributes about 6-8 per cent of total Indian rice production. Proportion of Basmati production is around 42 per cent and 22 per cent of total rice production in Haryana and Punjab respectively; while proportion is 8 per cent in Uttar Pradesh and 12 per cent in Madhya Pradesh. Haryana and Punjab contributes around 73 per cent of total Basmati production, 18 per cent comes from Uttar Pradesh, 6 per cent from Madhya Pradesh. Other states like Rajasthan, Uttaranchal, Himachal Pradesh and Jammu & Kashmir contributes 3 per cent of Indian Basmati production. (APEDA)

The total production of Basmati is 8773.78 thousand tonnes in 2014-15, in India that was improved or greater from the previous year 2013-14 production of basmati, which was 6616.13 thousand tonnes. Therefore, has been raised. Haryana is leading in the production of Basmati in 2014-15 by producing 3701.88 thousand tonnes. Punjab is at second number for producing 3498.88 thousand tonnes Basmati in 2014-15. Uttar Pradesh is at third rank in 2014-15 for producing basmati of 1260.69 thousand tonnes. Fourth rank 2014-15 is occupied by Jammu and Kashmir for producing 240.77 thousand tonnes of basmati. Fifth rank goes to Uttarakhand for the production of 66.41 thousand tonnes in 2014-15. In 2014-15, Himachal Pradesh is at sixth position for the production of 2.15 thousand tonnes of basmati and Delhi lies at seventh position for the production of 3 thousand tonnes of basmati in 2014-15.

A combination of factors were responsible for India to reach this pinnacle as one of the world's major rice exporter, which include the research efforts in developing suitable varieties and farm management practices which were largely adopted by the farmers. In addition, appropriate steps taken by the government to liberalize trade policy and licensing procedures so as to foster agricultural exports, the efforts made by the All India Rice Exporters Association (AIREA) & Agricultural Processed Food Products Export Development Authority (APEDA), a major role in promoting Indian Agri-exports and the zeal of the exporters in establishing themselves as reliable and dependable suppliers of basmati and non-basmati rice played.

❑❑❑

Chapter 2

Theory of Agri-Supply Chain Management

Supply Chain Management is a network of facilities that produce raw materials, transform them into intermediate goods and then final products, and deliver the products to customers through a distribution system. Supply chains are principally concerned with the flow of products and information between supply chain member organizations procurement of materials, transformation of materials into finished products, and distribution of those products to end customers. The Supply Chain Management Program integrates sectors from manufacturing operations, purchasing, transportation, and physical distribution into a unified program. Successful supply chain management, then co-ordinates and integrates all of these activities into a seamless process. It embraces and links all of the partners in the chain. In addition to the departments within the organization, these partners include vendors, carriers, third party companies, and information systems providers. *(B.B. Handel).*

Supply chain management is typically viewed to lie between fully vertically integrated firms, where the entire material flow is owned by a single firm and those, where each channel member operates independently. Therefore, co-ordination between the various players in the chain is key in its effective management. Cooper and Ellram (1993) compare supply chain management to a well-balanced and well-practiced relay team. Such a team is more competitive, when each player knows how to be positioned for the hand-off. The relationships are the strongest between players who directly pass the baton, but the entire team needs to make a co-ordinated effort to win the race.

The primary objective of Supply Chain Management is to fulfill customer demands through the most efficient use of resources, including distribution capacity, inventory, and labour. In theory, a supply chain seeks to match demand with supply and do so with the minimal inventory. Various aspects of optimizing the supply chain include liaising with suppliers to eliminate bottlenecks; sourcing strategically to strike a balance between lowest material cost and transportation, implementing just-in-time techniques to optimize manufacturing flow; maintaining the right mix and location of factories and warehouses to serve customer markets; and using location allocation, vehicle routing analysis, dynamic programming, and traditional logistics optimization to maximize the efficiency of distribution.

Agricultural Supply Chain Management

Supply chains are concerned with the flow of products and information between supply chain actors, procurement of materials, transformation of materials into finished products, and distribution of those products to ultimate end customers. In this era, information-driven, integrated supply chains are enabling organizations to reduce inventory and costs, add product value, extend resources, accelerate time to market, and retain customers.

The actual and real measure of supply chain success is how well activities coordinate across the supply chain to create value for consumers, while increasing the profitability of every link in the supply chain. In other words, supply chain management is the integrated process of producing value for the end user or ultimate consumer.

2.1 Food Supply Chain Networks

A processing-based and organized agri-supply chain functions as a part of a very complex network. Figure 1 depicts a generic supply chain at the organization level within the context of a complete supply-chain network. Each firm is positioned in a network layer and belongs to at least one supply chain, i.e. it usually has multiple (varying) suppliers and customers at the same time and over time.

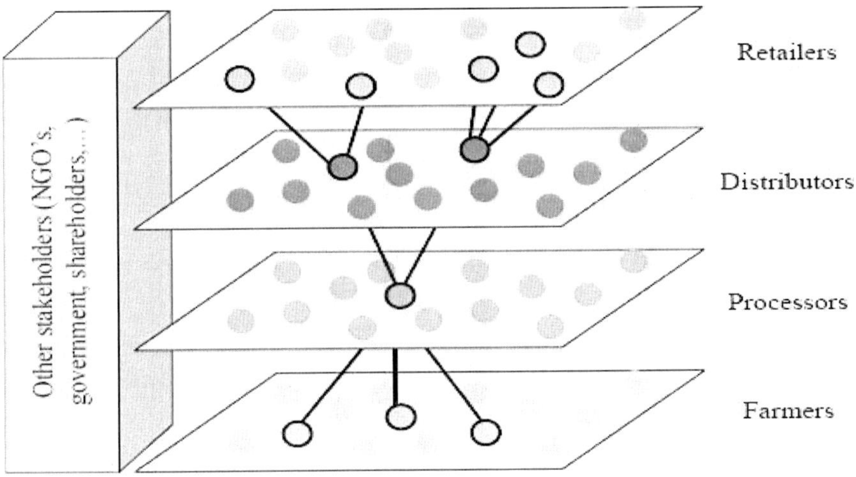

Schematic diagram of a supply chain from the perspective of the processor (bold flows) within the total FSCN (based on Lazzarini et al. 2001)

Source: Google and Lazzarini, 2001

2.2 The advantages for supply chain members

Individual suppliers, producers and marketers who are associated through a supply chain co-ordinate their value creating activities with one another and in the process, create greater value than they can, when they operate independently. Supply chains create synergies in one of three ways:

i) They expand traditional markets beyond their original boundaries and thus, increase sales volume for members;

ii) They reduce the delivered cost of products below the cost of competing chains and thus, increase the gross margin for the working capital committed by members of the chain ; and

iii) They target specific market segments with specific products and they differentiate the service, product quality or brand reputation of the products, they deliver to these market segments and thus, increase ease market contestability both at the producer end and at the consumer ends of the chain. At the consumer end, chains compete primarily through price, differentiated products and services and differentiated terms of sale. At the producer end of the chain, supply chains compete with one another primarily for "producer affiliation" and core vendor commitments.

2.3 Components of an Agri Supply Chain

Agri-business supply chain management (SCM) implies managing the relationships between the businesses responsible for the efficient production and supply of products from the farm level to the consumers to meet consumers' requirements reliably in terms of quantity, quality and price. In practice, this often includes the management of both horizontal and vertical alliances and the relationships and processes between the firms.

Agri-supply chains are economic systems which distribute benefits and apportion risks among participants. Thus, supply chains enforce internal mechanisms and develop chain wide incentives for assuring the timely performance of production and delivery commitments. They are linked and interconnected by virtue of shared information and reciprocal scheduling, product quality assurances and transaction volume commitments. Process linkages add value to agricultural products and require individual participants to co-ordinate their activities as a continuous improvement process. Costs incurred in one link in the chain are determined in significant measure by actions taken or not taken at other links in the chain. Extensive pre-planning and co-ordination are required up and down the entire chain to affect key control processes, such as forecasting, purchase scheduling, production and processing programming, sales promotion, and new market and product launches, etc. Following are the components of an organized agri- supply chain:

1. Procurement or sourcing
2. Logistic management
    - Transportation
    - Material management
    - On the premise of supplying mostly from production net stock
    - Warehousing
    - Logistics Network modelling
3. Organizational management
    a) Contracting
    b) Strategic alliances and partnerships

c) Vertical integration

i) Long term storage

ii) Packaging technology

iii) Cold chain management.

iv) Energy efficient transport

v) Quality and safety

4. Application of Efficient Consumer Response (ECR) System

- Electronic scanning of price and product at the point of sale
- Streamline the entire distribution chain

In terms of management practices, Basmati rice is the basic grain of daily consumption in India, it always has the significant demand which shows a seasonal growing trend, especially at the time of festivals and during the seasons of cultural events. To identify the appropriate demand and meet them properly in a profitable way is primary concern of basmati rice supply chain. A proper supply chain management framework is very essential for efficient sourcing, processing, distribution, and retailing and hence, meeting the customer demands without facing a situation of lost sales. Production and business of rice has been one of the most traditional and major concerns of Indian Economy, but still no proper supply chain framework for it has been developed, which very often causes unfulfilled demands, stock out and overstocking, and distribution issues.

2.3.1 Decision of Supply Chain

We classify the decisions for supply chain management are classified into two broad categories viz., strategic and operational. As the term implies, strategic decisions are made typically over a longer time horizon. These are closely linked to the corporate strategy and guide supply chain policies from a design perspective. On the other hand, operational decisions are short term, and focus on activities over a day-to-day basis. The effort in these types of decisions is to effectively and efficiently manage the product flow in the "strategically" planned supply chain.

There are four major decision areas in supply chain management: Location, Production, Inventory, and Transportation (distribution), and there are both strategic and operational elements in each of these decision areas.

2.3.2 Location Decisions

The geographic placement of production facilities, stocking points, and sourcing points is the natural first step in creating a supply chain. The location of facilities involve a commitment of resources to a long-term plan. Once the size, number, and location of these are determined, so are the possible paths by which the product flows through to the final customer. These decisions are of great significance to a firm since they represent the basic strategy for accessing customer markets, and will have a considerable impact on revenue, cost, and level of service.

2.3.3 Production Decisions

The strategic decisions include what products to produce, and which plants to produce them in, allocation of suppliers to plants, plants to DC's, and DC's to customer markets. As before, these decisions have a big impact on the revenues, costs and customer service levels of the firm. These decisions assume the existence of the facilities, but determine the exact path(s) through which a product flows to and from these facilities.

2.3.4 Inventory Decisions

These refer to means by which inventories are managed. Inventories exist at every stage of the supply chain as either raw material, semi-finished or finished goods. They can also be in process between locations. Their primary purpose is to buffer stock against any uncertainty that might exist in the supply chain. Since holding of inventories can cost anywhere between 20 to 40 per cent of their value, their efficient management is critical in supply chain operations.

2.3.5 Transportation Decisions

The mode choice aspects of these decisions are the more strategic ones. These are closely linked to the inventory decisions, since the best choice of mode is often found by trading-off the cost of using the particular mode of transport with the indirect cost of inventory associated with that mode. While air shipments may be fast, reliable, and warrant lesser safety stocks, they are expensive. Meanwhile shipping by sea or rail may be much cheaper, but they necessitate holding relatively large amounts of inventory to buffer against the inherent uncertainty associated with them. Therefore, customer service levels and geographic location play vital roles in such decisions.

2.4 Importance of Supply Chain Management

Supply Chain Management (SCM) is an essential element to operational efficiency. Supply Chain Management can be applied to customer satisfaction and company success as well as within societal settings, including medical missions, disaster relief operations and other kinds of emergencies, Cultural Revolution, and it can help improve quality of life. Because of the vital role, Supply Chain Management plays within organizations, employees seek employees with an abundance of Supply Chain Management skills and knowledge. Supply Chain Management is critical to business operations and success. Supply Chain Management is helpful in reducing costs. It provides better medium for information sharing between partners. It is also helpful in improving customer's satisfactions as well as services. Supply Chain Management os also helpful in increasing cashflow and improves quality and gives higher profit margin.

2.5 Review and study related to Resource structure of Selected Basmati Rice mill Unit

Hofstede (1980) stated that organizational culture refers to the collective programming of the mind that distinguishes the members of one organization from another. This includes shared beliefs, values and practice that distinguish one organization from another. The beginning of formal writing in an organizational culture started with Petigrew (1979). He introduced the anthropologist concepts like "symbolism, myths," and "rituals" that could be used in organizational analysis.

Sekaran (1981) in his study used a multi-variate cross-cultural approach to explore the meaning of two attitudinal concepts - job involvement and job satisfaction. His sample consisted of 267 white collar workers from US banks and 307 from Indian banks. He found that in both the cultures, job variety and stress were the two common predictors for job satisfaction. Income was a third significant predictor in the US, while communication was the third additional predictor in India. For job involvement, age appeared to be a differential predictor. She concludes from her study that a manager should concentrate on job design and stress reduction to enhance the job satisfaction of employees in both cultures.

Sinha and Sahaya (1981) observed that even the best sophisticated and modern equipment could give no result, if the human machinery marshalled to operate them was not sincere, effective and efficient in using them to the best advantage of the organization. So, he urged the personnel management to create conditions, under which each individual employee would not only give maximum job performance, but would also feel maximum job satisfaction.

Lawler (1982) defined Quality of Work Life in terms of job characteristics and working conditions. He highlights that the core dimension of the entire Quality of Work Life in the organization is to improve employees, well-being and productivity. The most common interaction that relates to improvement of employees, well-being and productivity is the design of the job. The job design that is able to provide higher employee satisfaction is expected to be more productive.

Buchanan and Boddy (1983) in their study discussed that human force is more important than economic market forces. It is, therefore, important that work is organized to develop skills and motivation for organizational effectiveness.

Singh (1983) in his study of managers from the public sector concludes that the overall perceived Quality of Work Life in the Indian industries is considerably poor. While this finding is common across all the work dimensions studied; the Quality of Work Life is perceived to be the poorest in the area of democratization of work, culture leading to a state of mismatch between motivational and the existing Quality of Work Life. He found that Quality of Work Life was perceived to be poorer by private sector employees in comparison to public sector employees. Quality of Work Life was operationalised for the study of human growth, such as exciting work place, creativity and innovativeness concern for people and democratization of the work process.

Hartenstein and Huddleston (1984) emphasized that for Quality of Work Life measures to be successful, management and labour must have shared values. Without such shared values, managers are often authoritarian and deny workers sense of involvement, responsibility and autonomy, resulting in a lack of commitment and low productivity.

Mirvis and Lawler (1984) suggested that Quality of Work Life was associated with satisfaction with wages, hours and working conditions. He described the "basic elements of a good quality of work life", such as safe work environment, equitable wages, equal employment opportunities and opportunities for advancement.

Nitish De (1984), in his study, discussed about concepts, problems, issues and experiences in quality of work life. He initiated work on job or work designing in India through his associates. Successful experiments have been carried out on several settings by them. His major findings indicate that to make the quality of work life better, the practice of job redesign and work commitment are suggestible.

Michael Maccoby (1984) identified the job satisfaction of employees and supervisory of Bell System employees over a 5-years period. He found that the employees and supervisory staffs were satisfied with pay and benefits and motivated to work productively, but they were dissatisfied with technology and perceived too much supervisory control. They believed that they were mismanaged, pushed around, not listened to, and that the spirit of service was being eroded by the drive to increase profit.

Venkatraman & Ramanujam (1986) analyzed that organization performance has been the most important issue for every organization, be it profit or non-profit one. It has been very important for managers to know which factors influence an organization's performance in order for them to take appropriate steps to initiate them. However, defining, conceptualizing, and measuring performance have not been an easy task. Researchers among themselves have different opinions and definitions of performance, which remains to be a contentious issue among organizational researchers (Barney, 1997). The central issue concerns with the appropriateness of various approaches to the concept utilization and measurement of organizational performance.

Mathur and Paranjpe (1988), in his study, "Identification of training areas for special educators as related to job satisfaction factors of Quality of Work Life perceptions" found significant relation of overall job satisfaction with Quality of Work Life feelings and condition.

Baba and Jamal (1991) studied various indicators of quality of working life, including job satisfaction, job involvement, work role ambiguity, work role conflict, work role overload, job stress, organizational commitment and turn-over intentions. Baba and Jamal also explored routinization of job content, suggesting that this facet should be investigated as part of the concept of quality of working life.

Asit Narayan and Amarnath Jha (1992) stated that labour is human. Employees differ in mental abilities, emotional stabilities, institutions and sentiments. They also stated that "Quality of Work Life is a generic phrase that covers a person's feeling about every dimension of work including economic rewards and benefits, safe and healthy working conditions, organizational and interpersonal relationship and its intrinsic meaning in a person's life".

Bino Thomas (1995), in his study stated that for the sustaining of quality circle activities, the employees must come to believe that their support and participation will benefit themselves as well as the organization. Employees should not feel that quality circle is just another management gimmick to use them. They must be convinced that quality circle is a people-building philosophy rather than a people-using approach. Suggestions and recommendations of the work related problems flow upwards, in lieu of a top-down approach with instructions and directions flowing from the top downwards. Developing a positive organizational climate with minimum stress and strain is a sure way for improving productivity in any organization. For the quality circle to be enduring, it should be an integral part of the total quality of work life in the organization.

Srivastava (1996) pointed out that organizational climate and higher order needs (self-esteem, autonomy and self-actualization) are found to be positively related to job involvement. In is necessary to point out that this study has not noted the variables, which had been undertaken for study as the term Quality of Work Life but organizational climate, higher order needs and all other bio-social needs are the determinants of Quality of Work Life.

Heskett and Schlesinger (1997) stated that Quality of Work Life is the feelings that employees have towards their jobs, colleagues and organizations that ignite a chain leading to the organizations growth and profitability. A good feeling towards their job means the employees feel happy doing work which will lead to a productive work environment. This definition provides an insight that the satisfying work environment is considered to provide better Quality of Work Life.

Joseph Zakhariya P.J (1999), in his study stated that job inherent factors indicated that job security was uppermost in the minds of all employees, irrespective of officers and workmen. Analysis of the factors influencing quality of work life reveals that factors like age, experience, educational qualification, etc. bear no relation to quality of work life. However, factors like job satisfaction and job perception in terms of suitability and challenges offered were directly related to quality of work life. This was more prominent in the case of workmen who may be attributed to the monotonous nature of jobs and lesser scope for promotion.

Duncan Gallie (2003) studied and stated that there is a comparison of employees, perceptions of the quality of working tasks, the degree of involvement in decision-making, career opportunities, and job security to see whether the Scandinavian countries have a distinctive pattern from other European Union countries. It was found that on the aspects of working life that were most central to the reform programmes, the results were consistent with the view that there could be societal effects deriving from the policy orientations of the major economic interest groups.

Wilson (2003), in his study observed that workers were generally satisfied with the environment in which they worked. The satisfaction rate was high among the public sector workers than the private sector workers. Regarding style of supervision, both the categories were satisfied. Both the categories were not satisfied with respect to promotion and involvement in decision-making. He also found that public sector employees had more satisfaction than private sector employees with respect to reward, human relations, behaviour of co-workers and nature of job.

Wisner (2003) in supply chain management strategies, supplier relationship activities play an important role. Long-term relationships refer to intention that the arrangement is not going to be temporary (Chen and Paulraj, 2004). Through close relationship supply chain partners are willing to share risks and reward, and maintain the relationship on long-term basis (Landeros and Monczka, 1989; Cooper and Ellram, 1993; Stuart, 1993; Thatte, 2007). Toni and Nassimbeni (1999) identified that a long-term perspective between the buyer and supplier increase the intensity of firm-supplier integration. Firms that integrate with customers including planning, implementing, and evaluating a successful relationship between the provider and recipient of both upstream and downstream of the supply chain. Therefore, customer relationship management (CRM) is not only focused on inbound customer relationships but also on outbound customer relationships in Supply Chain Management. Customer relations related to the company's ability to communicate to the delivery of appropriate products and services to customers locally and globally in the right time, right place, and appropriate of quantity and quality. Customer linkage, especially sharing product information with customers, receiving customer orders, interact with customers to manage demand, after placing the order system, share the status of orders with customers on scheduling orders, and product delivery stage (Lee, *et al.*, 2007).

Chen and Paulraj (2004) also conduct the research regarding supply chain management practices, they investigated long-term relationship, cross-functional teams, supplier base reduction, and supplier involvement. The same with Chen and Paulraj, Min and Mentzer (2004) also examined in their study long-term relationship, information sharing, co-operation process integration and supply chain leadership, underlying the supply chain management practices. Lie *et al.* (2005, 2006); Thatte (2007) identified supply chain management practices in the form of strategic supplier partnership, customer relationship, and information sharing. This research adopts the same supply chain management practices (supplier partnership, customer relationship and information sharing). However, this study conducted in Malaysia perspective, especially in consumer goods industry. Li et al (2005); Thatte (2007) have developed a valid and reliable instrument to measure supply chain management practices. The similar instrument also adopted in this research. The literature also depicts supply chain management practices from different perspectives with goal of improving competitive advantage of firm. By improving competitive advantage of the firm, organization could improve its performance. Three dimensions of supply chain management practices lead to supply chain responsiveness.

Li et al, (2006) in their study observed that the supply chain management practices that encompasses supplier partnership, customer relationship and information sharing, supply chain responsiveness that includes operation system responsiveness, logistic process responsiveness, and supplier network responsiveness, competitive advantage of the firm. Supply chain management practices as a multi-dimensional construct that encompasses upstream and downstream sides of supply chain. Donlon (1996) stated that outsourcing, supplier partnership, information sharing, cycle time, compression and continuous process flow, as a part of supply chain management practices. While Tan *et al.* (1998) represented supply chain management practices in the form of quality, purchasing, and customer relationship. Alvarado and Kotzab, in their empirical study focused on supply chain management practices on inter-organizational system used, core competences, and elimination of excess in inventory through postponement. The key aspect of supply chain management practices according to Tan *et al.* (2002) were supply chain integration, information sharing, customer service management, geographic proximity, and JIT capabilities. Lee (2004) focused on five practices at supply chain level that are a key to create supply chain responsiveness. They include outsourcing, strategic supplier partnerships, customer relationship, information sharing, and product modularity.

Armstrong and Stephens (2008) stated that organizational structures can be classified as unitary, divisionalised, centralized, matrix, and process. Unitary structure, the most common structure, can be described as single and separate unit without any divisions, in which the heads of each major function directly reports to the top. However, such key functions may vary from one organization to another.

2.6 Review of study related to Supply Chain of Basmati Rice and its Constraints

Cohen and Lee (1985) developed a conceptual framework for manufacturing strategy analysis, where they describe a series of stochastic sub-models, that considers annualized product flows from raw material vendors via intermediate plants and distribution echelons to the final customers. They use heuristic methods to link and optimize these sub- models. They later give an integrated and readable exposition of their models and methods in Cohen and Lee (1988).

Porter, M. E. (1985) stated that the supply chain, also known as value chain is a concept from business management that was first described and popularized by Michael Porter in his book, Competitive Advantage: Creating and Sustaining Superior Performance.

Cohen and Lee (1989), in their study presented a normative model for resource deployment in a global manufacturing and distribution network. Global after-tax profit (profit-local taxes) is maximized through the design of facility network and control of material flows within the network. The cost structure consists of variable and fixed costs for material procurement, production, distribution and transportation. They validate the model by applying it to analyze the global manufacturing strategies of a personal computer manufacturer.

Powell (1990) stated that inter-organizational supply network can be acknowledged as a new form of organization. However, with the complicated interactions among the players, the network structure fits neither "market" nor "hierarchy" categories. It is not clear what kind of performance impacts different supply network structures could have on firms, and little is known about the co-ordination conditions and trade-offs that may exist among the players.

Scott and Westbrook (1991), in their study stated that a firm's customer relationship practices can generate the organizational success in supply chain management practices efforts as well as its performance (Ellram, 1991; Turner, 1993). The success of supply chain management encompasses customer integration at the downstream and supplier integration at the upstream, considering that each entity in a supply chain is a supplier as well as a customer (Tan et al., 1999; Thatte, 2007). In the competitive business, better relationship management with customers is crucial for organization success (Wines, 1996). Good relationship with business partners, including key customers are important role to success of supply chain management practised by organization (Moberg et al, 2002; Tathee, 2007). Customer relationship recognized as an internal component of an organization's market strategy to increase sales and profits (Bommer et, 2001; Thatte, 2007). Close customer relationship allow product differentiation from competitors, help sustain customer satisfaction and loyalty, and elevated the value provide to customer (Margaretta, 198; Thatte, 2007).

Christopher (1992) stated that a supply chain is a network of organizations to perform a variety of processes and activities to generate value in the form of products and services to end consumers. Supply Chain Management involves an integrated and process-oriented approach to the management, design and control of the supply chain, with the aim of producing value for the end consumer, by both customer service and reduces cost (Bowersox and Closs, 1996; Giannoccaro and Pontrandolfo, 2002). The supply chain management has dual purpose, in one side is to improve the performance of an individual organization as well as that of the entire supply chain. In other side, the supply chain management reduces organization total cost (Li et al, 2006). Supply Chain Management (SCM) is the discipline which is relatively new and lack of theoretical conceptual framework of established or accepted methodology, in general. Lambert, *et al.*, (1998) observed that "the term of Supply Chain Management was originally introduced by consultants in the early 1980s. Since the early 1990s, scholars have attempted to give structure to the development of SCM. Although the short story of this field, but the SCM literature has grown rapidly (Larson and Rogers, 1998). Shapiro (2001) showed that the SCM combining concepts from different disciplines, such as strategic management and theory of the formation of the company; logistics, production and inventory management; accounting management; scientific forecasting and marketing, and operations research.

Bowersox and Closs (1996) showed that to be fully effective in current's competitive business, companies must develop their integrated behavior to incorporated customers and suppliers. This expansion of integrated behaviors, through external integration, referred to by Bowersox and Closs (1996) as supply chain management. The philosophy of supply chain management turns into the implementation of SCM: a set of activities that carries out the philosophy. The set of activities are coordinated effort called supply chain management between the supply chain partners, such as suppliers, manufactures and customers (Greene, 1991). Thatte (2007) stated that strategic supplier partnership as the long-term relationship between the organization and its supplier.

Gunasekaran *et al.* (2001) asserted that a strategic partnership emphasizes long-term relationship between trading partners and promote mutual planning an problem solving efforts. Strategic partnership between organizations promote shared benefits and ongoing collaboration in key strategic areas like technology, products, and market (Yoshino and Rangan, 1995; Thatte, 2007). Strategic partnerships with suppliers lead organization working closely and effectively with a few suppliers rather than many supplier that have been selected on the basis of cost efficient. Many advantage of consisting supplier early in the product-design process are that suppliers can offer cost effective design alternative, assist in selecting better components and technologies, and aid in designing assessment (Tan *et al.*, 2002; Thatte, 2007).

Harland (1996) studied and stated that Supply Chain Management (SCM) is the management of a network of interconnected businesses involved in the ultimate provision of product and service packages required by end customers. Supply chain management spans all movement and storage of raw materials, work-in-process inventory, and finished goods from point of origin to point of consumption.

Coase (1998), in his study stated that in 21st century, changes in the business environment have contributed to the development of supply chain network. First, as an outcome of globalization and the proliferation of multinational companies, joint ventures, strategic alliances and business partnerships, significant success factors were identified, complementing the earlier "Just-In-Time", "Lean Manufacturing" and "Agile Manufacturing" practices. Second, technological changes, particularly the dramatic fall in information communication costs, which are a significant component of transaction costs, have led to changes in co-ordination among the members of the supply chain network.

Peter Drucker's (1998) studies and revealed that Organizations increasingly find that they must rely on effective supply chains or networks, to compete in the global market and networked economy. In this concept of business relationships, extends beyond traditional enterprise boundaries and seeks to organize entire business processes throughout a value chain of multiple companies.

Raphael Kaplinsky and Mike Morris (2000) defined the Supply chain and value chain as the full range of activities, which are required to bring a product or service from conception, through the different phases of production (involving a combination of physical transformation and the input of various producer services), delivery to final consumers, and final disposal after use. The definition recognizes the flow of services as well as products in value chains.

Whipple & Frankel (2000) revealed that Supply chain management requires a process having the attributes of long and complex. The successful implementation depends on how far one has gained insights over the concept. The aspect of execution also depends on how far the relationship is fixed out involving individuals, organizations, and supply chain networks (Spekman et al., 1998). Hence, the need is to arrange the attributes of the organization and its culture in new perspective (Neely, 1998), where the need is to have more commitment, resources and investment of time to develop. Since, so many things are involved, this is recognized as a multi-factor process, where the dependency is on close and long-term relationships not only within the organization, but also other involved organizations as well (Saad et al., 2002). The success of it depends on factors like on shared learning, greater transparency, and trust while dealing with the challenging aspect of development of a new culture. The demanding aspect of supply chain management is to maintain and perform well among the competitors on continuous basis, for which the emphasis should be on co-ordination and integration amidst of heavy reliance on suppliers and the advantage of outsourcing. Hence, it becomes important to have the concept of supply chain management with the capacity for improvement on permanent basis along with performance dimension.

Ackerman (2001) stated that researchers have recognized these kinds of supply network structures as a new organization form, using terms, such as "Keiretsu", "Extended Enterprise", "Virtual Corporation", "Global Production Network", and "Next Generation Manufacturing System". In general, such a structure can be defined as "a group of semi-independent organizations, each with their capabilities, which collaborate in ever-changing constellations to serve one or more markets in order to achieve some business goal specific to that collaboration".

Mentzer *et al.*,(2001a) stated that Supply chain is a set of Organizations directly linked by one or more of the upstream and downstream flows of products, services, finances, and information from a source to a customer. Managing a supply chain is 'supply chain management'.

Mentzer *et al.*,(2001b) stated that Supply Chain Management is the systematic, strategic co-ordination of the traditional business functions and the tactics across these business functions within a particular company and across businesses within the supply chain, for the purposes of improving the long-term performance of the individual companies and the supply chain as a whole.

Stern et al., (2001) studied and stated that Supply chain is the set of entities that collectively manufactures a product and sells it to an endpoint. SCM practices have been defined as a set of activities carried out to promote efficient management of its SC (Li, Raghunathan, Raghunathan and Rao, 2006).Although, there is a few consensus with respect to the understanding and definition of SCM (Mentzer *et al.*, 2001), Burgess *et al.*, 2006, Stock & Boyer, 2009 & Stock *et al.*, 2010).

Agarwal and Shankar (2002) proposed an analytic network process based model for analyzing the alternatives affecting supply chain performance. It also provides the decision methodology to prioritize these alternatives, so that supply chain performance can be improved. A process-based systematic perspective was employed to build an effective model to measure the holistic performance of complex supply chains. Fuzzy set theory was introduced to address the real situation in judgement and evaluation processes. Fynes at al. (2005), Gunasekaran *et al.* (2001) developed a conceptual framework to link up SC relationship dynamics and manufacturing performance and found out the less evidence of SC relationship dynamics on manufacturing performance.

Ho *et al.* (2002), in his study highlighted major conceptual gaps in their research. They discussed the need to operationalise and model the SCM constructs, so they might be viewed from a business process perspective with better performance as a desired objective.

Simatupang and Sridharan, (2002) defined information sharing as the access to private data between business partners, thus enabling them to monitor the progress of products and orders, as they pass through various processes in the supply chain. They identified some of element that comprise information sharing, consisting data acquisition, processing, storage, presentation, retrieval, and broadcasting of demand and forecast data, inventory status and location, order status, cost-related data, and performance status. They also add that information sharing pertaining to key performance metric and process data improves the supply chain visibility, thus enabling effective decision-making. Information shared in a supply chain is of use only if it is relevant, accurate, timely, and reliable (Simatupang and Sridharan, 2005; Tathee, 2007). Information sharing with business partners enables organizations making better decisions and making action on the basis of greater visibility (Davenport, et al, 2001; Tathee, 2007). Lumnus and Vokurka (1999, cited in Thatte, 2007) stated that in order to make the supply chain competitive, a necessary first step is to acquire a clear understanding of supply chain concepts and be willing to openly share information with supply chain partners. In business competitive world now-a-days, business organization should develop their supply chain in order to get customer responses.

Tan (2002) analyzed in order to explore the trends of supply chain management practices and issues, conducted a survey with respondents belonging to supply and materials management professionals from United States working at higher echelons. Moreover, this study encapsulate about the performance in context to the practices and concerns through the statistical edge of bi-variant correlation and multiple linear regression analysis. The usual notion was that performance depended on relevant supply chain management practices. Study by Li *et al.* (2005), focused on six dimensions of SCM practices (strategic supplier partnership, customer relationship, information sharing, information quality, internal lean practices, and postponement). The study was done to obtain to measure the overall supply chain performance.

Quayle (2003), in his study reflected about the benefits of a supply chain management (SCM) practice and how it required to be implemented in the trends prevalent in small- to medium-sized enterprises operating out of UK. The analysis recognized about the necessity of implementing SCM techniques and how it matters to customers and smaller suppliers. The outcomes comprising of survey conducted among 288 firms, which reflected about not resorting to adaption in right perspective right from the traditional time to the modern collaborative "e" - supply chain; moreover, the focus was to deal with the bottlenecks of supply chains so as to have better performance and stand apart from the crowd by leveraging the benefits of SCM.

Sahay and Mohan (2003) made attempted to explore on the supply chain practices being implemented by the Indian industry and provided suggestive measures need to be taken for better result. A combined survey conducted by Management Development Institute, Gurgaon and KPMG India, involving 156 organizations. Here, the highlight of the survey was on the stand of four major supply chain dimensions. As per the recommendation of that paper, Indian industry needs to put supply chain strategy on the lines of business strategy and integrate the processes for supply chain integration. Moreover, it was suggested to develop partnerships for reducing inventory and emphasize on infrastructure and implementation of technology to develop a supply chain conducive to India.

Chin *et al.* (2004) conducted a study and reflected about the outcomes related to components responsible for success in generating and adopting strategies of supply chain management in context to Hong Kong manufacturers. Moreover, the results would assist in facilitating more insights over the success factors for implementing Supply Chain Management strategies effectively to minimize supply chain costs and to raise the bar of serving customer.

Hines (2004a) observed a problem frequently observed in unmanaged supply chains in the Flour Mills as the bullwhip effect. This effect is an oscillation in the supply chain caused by demand variability. This problem must be addressed in order to avoid the poorer service and higher costs that stem from it. An unmanaged supply chain is not inherently stable. Demand variability increases, as one moves up the supply chain away from the retail customer, and small changes in consumer demand can result in large variations in orders placed upstream. Eventually, the network can oscillate in very large swings, as each organization in the supply chain seeks to solve the problem from its own perspective. This phenomenon is known as the bullwhip effect and has been observed across most industries, resulting in increased cost and poorer service.

Hines (2004b) analyzed that "Supply chain strategies require a total systems view of the linkages in the chain that work together efficiently to create customer satisfaction at the end point of delivery to the consumer. As a consequence, costs must be lowered throughout the chain by driving out unnecessary costs and focusing attention on adding value. Throughout efficiency must be increased, bottlenecks removed and performance measurement must focus on total systems efficiency and equitable reward distribution to those in the supply chain adding value. The supply chain system must be responsive to customer requirements."

Li *et al.*, (2005) in their study, Conducted an empirical survey to confirm six different dimensions of SCM practices relating to strategic supplier partnership, customer relationship, information sharing, information quality, internal lean practices, and postponement.

Power (2005) presented a conceptual framework to investigate supplier relations, lean manufacturing, environmental management practices; and their relationship to one another. It is found that efforts to improve a supplier's environmental management practice raise critical issues of transaction costs and efficacy of approach for the buyer. It is recommended that an environmental bias is to be introduced into the decision-making process, which would allow more environmentally conscious decisions to be made.

Wong *et al.* (2005) in their study assessed SCM practices, and narrowed down the gaps lying in context to practical and theoretical as far as toy supply chains are concerned. Study conducted in the past involving a longitudinal and in-depth case study found place in this paper having relevance to international toy manufacturer, containing questionnaire for conducting qualitative semi-structured interviews among 11 main European toy retailers. As per the study conclusion, three prominent SCM practices were essential for toy retailers, having the behaviour order of (one-off, JIT, and mixed model), and relevance of one SCM practice for toy manufacturers (traditional mass-production or push-models). Gradual knowledge diffusion was not the sole responsible factor for minimal responsive practices prevalent in the toy supply chain. SCM technology is not at par to manage the aspect of volatility and confined to certain period. Hence, the need for proposal was sought for to deal with theoretical gaps and development of new theories.

Lia et al., (2006) developed and conceptualized the Supply Chain Management practices in Five- dimensions, such as relationship with customer, partnership with strategic supplier, information sharing level, information sharing quality and postponement) and also identified the association between Supply Chain Management practices, competitive advantage, and organizational performance. From about 196 organizations, data was collected and finally structural equation modelling was used in testing the framework. The final analysis of this study resulted in proving that enhanced competitive advantage and organizational performance can be achieved by higher level of Supply Chain Management. In fact competitive advantage is directly linked with a positive impact on organizational performance.

Chopra and Meindl (2007) studied that all the stages which directly or indirectly meets the need of customer comes under the purview of supply chain. In addition to manufacturers and suppliers, the supply chain also takes into cognizance of the importance of transporters, warehouses, retailers, and customers themselves. Again, supply chain encompasses the functions, which are relevant to meet the customer requirements; it is applicable for all organizations. However, it is not confined to development of new product, marketing, operations, finance, distribution, and customer service. Therefore, the need is to manage the interfaces cautiously on the part of all the organizations as a part of supply chain management (Von Hipple, 1986), and there is a need for streamlining the processes of upstream and downstream (Christopher & Juttner, 2000). To have a healthy development and maintain rapport with buyers and sellers over a longer period of time, the focus should be on co-ordination and integration (Kosela, 1999).

Gruen and Corsten (2007) studied that non-availability of stock would lead to loss of $200,000 in the lifetime lost sales. In addition, the findings also emphasized that poor stocking affected the brand loyalty and short-term sales. Thus, stock level planning plays an important role in Supply Chain Management efficiency.

Koprulu and Albayrakoglu (2007), in their study examined the main factors for a successful supply chain in a globalised scenario. An analytical hierarchy process (AHP) model that could be used by an apparel a company for selecting the suppliers was presented and a strategy of supplier relationship management (SRM) was formulated with the results of the model as the basis. Additionally, strategic priorities for the problem of supplier selection were identified leading to development of weights for the selection of the right supplier that suits the strategy of the company. It was shown that the most significant strategic priorities are the trust delivery and quality, which are to be considered in the problem of supplier selection for matching category, that represents >65% of the total weight. Therefore, a practical tool that assists the stakeholders to get a logical idea in the selection of the best vendor, accounts all the business aspects which are to be in alignment to the goals and strategies of the company. On using this tool regularly, aids in the inspection of the plans and also ensures nil deviation from the goals set.

Petrovic Lazarevic *et al.* (2007) identified five aspects of SCM practices training, information technology (IT), information sharing, internal operations and supplier and customer relations. The Australian manufacturing industry survey was conducted to learn the level to which the practice of SCM was implemented. It analysed the relationship between SCM practices and the performance of the firm. The results mainly emphasized the contribution and sharing of information through proper use of IT to the better operational performance of the firm. It was also found in the study that significant training on IT is required to promote the most performance measures.

Shergill (2007) found that production of rice in India is questionable on economic and ecological grounds. The decline is the concern of study and it can be either due to the technological obsolescence, the production, distribution, and the issues of market, or all of these issues. All these issues form it as a concern of the supply chain management, which is defined as a philosophy, which study the entire flow of material and goods throughout the channel in an integrated way, starting from supplier to the end customer. It is an integration of three or more establishments, whose activities are interlinked together directly with a purpose to flow the products, services, finances and information either in upstream or downstream way between the end customer and the source . One can say that for proper functioning of the rice sector in the agriculture industry, and regaining its upward growing trend, it is necessary to resolve its supply chain problems, and hence, a complete study of the rice supply chain management and its re-engineering is essential.

Thatte (2007) in his study stated that the sub-constructs for supply chain responsiveness include operation system responsiveness, logistic process responsiveness and supplier network responsiveness. Operation system responsiveness is the ability of firm's manufacturing system to address changes in customer demand. It includes both manufacturing and service operation. Duclos *et al.* (2003); Lumnus *et al.* (2003) emphasize that responsiveness at each company of the chain is an integral component of supply chain responsiveness. Logistic process responsiveness is the ability of company's outbound transformation, distribution and warehousing system to address changes in customer demand. Fawcett (1992) stated that the responsive in logistic process is a crucial component in the supply of a responsive supply chain strategy. Logistics and distribution management encompasses the transformation activities of goods from suppliers to manufacturer to distribution centers to final point of end users (Ricker and Kalakota, 1999; Duclos *et al.*, 2003; Thatte, 2007).

Johnson *et al.* (2008) revealed and stated that the performance of a supply chain is measured in terms of profit, average product fill rate, response time, and capacity utilization.

Lambert (2008) studied and stated that Global supply chain forum sees SCM as the integration of key business processes across the supply chain for the purpose of creating value for customers and stakeholders. A supply chain is a network that includes vendors of raw materials, plants that transform those materials into useful products, and distribution centers to get those products to customers (Carlson, 2005). Without any specific effort to coordinate the overall supply chain system, each organization in the network has its own agenda and operates independently from the others. However, such an unmanaged network results in inefficiencies. For example, a plant may have the goal of maximizing throughput in order to lower unit costs. If the end demand seen by the distribution system does not consume this throughput, there will be an accumulation of inventory. Clearly, there is much to be gained by managing the supply chain network to improve its performance and efficiency (Collins, 2003).

Anonymous (2009) defined Supply Chain Management as "SCM surrounds all the logistics management and planning activities (sourcing, procurement and conversion). Specifically, it also includes collaboration and co-ordination with channel partners who could be customers, intermediaries, suppliers and third party service providers. Basically, SCM combines demand and supply management inside and across the firms". The major task of SCM is an integrating function with chief business functions and processes inside and across the companies into a cohesive and high-performing business model. This responsibility includes all the logistics management activities noted above and manufacturing operations, which drives co-ordination of processes and activities with and across marketing, sales, product design, finance and information technology.

Ganesan *et al.* (2009) discussed how recent trends and changes in retailer supply chain practices influenced retailer's performance. Moreover, they provided remarks on global sourcing practices, multichannel routes to market and relationship-based innovation. The first trend concerned a move towards sourcing practices on a worldwide basis and included the rising need for corporate social responsibility, various country-of-origin concerns and some best practices for electronic procurement activities. Again they stated that the second trend was the disaggregation and innovation that arose from employing a multichannel route to market and the delicate power balance and conflict management needed that result. Lastly, it was considered that how the nature of inter-firm was tied between retailers and their organizational partners might better facilitate either product or process innovations among players, looking at how changes in technology and the structural characteristics of the organization impeded or facilitated attempts to advance joint efficiency.

Plazibat and Brajevic (2009), in their study identified how retailers who understood the SCM functions and wanted to be successful should utilize abilities of all members of their supply chain and tended to be more efficient and profitable. Therefore, findings indicated that focus of supply chain management should be on management of relationships and mutual trust of all members involved in supply chain in order to achieve the best market results. Relationships tied companies to each other and directly had impact on the success. Finally, retailers were recognized as vital part in satisfying customers demands by themselves and in co-operation with all members of supply chain.

Amaresha *et al.* (2010) discovered the significance of logistics and supply chain management in organised retail markets and its impact and to find out the problem areas in supply chain management on organised retail store (Critical to customer, critical quality and voice of the customer). In the organised retail market in India, the position of supply chain is very vital as the Indian customers demand a selection of product mix at affordable prices and it is ensured to the customers in all the variety of offering that company decides for its customer, be it cost, service, or the quickness in responding to ever changing taste of the customer.

Madhavedi and Acharyulu (2010) focused on the traditional practices in supply chain of fresh produce in India. The responses were taken from vital links in the downstream of the supply chain. Thus, the hypothesis was tested on delivery delays, rate of perish ability and replenishment times and ANOVA techniques were used to analysis. There commendations from that analysis were to improve the performance of the fresh produce supply chain. It could be concluded that there was a wide scope for improvement of supply chain practices in the fresh produce.

Singh *et al.*, (2010) conceptualized and developed five secondary constructs for supply chain practices (use of technology, SC speed, Customer satisfaction, SC integration, and Inventory management). Four primary competitive advantage constructs (Inventory Management, Customer Satisfaction, Profitability, and Customer Base Identification) and six primary organizational performances construct (Market Performance, SC Competencies, Stakeholder Satisfaction, and Innovation and Learning) were identified in this research. The data was collected from 10 non- livestock organised and the relationships among supply chain practices, competitive advantage, and organizational performance, were tested in the proposed framework using structural equation modelling. Finally, the study looked on the competitive advantages of Indian retailers with focus on supply chain planning The results indicated that Indian retailers knew that competitive advantage had high impact.

Sukati *et al.*, (2011) explored the relationship between supply chain management strategy and practices on supply chain performance. The data was collected from different managers belonging to functions like purchasing, Supply Chain Management, transportation, materials, operations, etc. from different industries. Convenience sampling was used to collect data and standard deviation, correlation and regression was used. The finding showed that supply chain management practices had a significant relationship with supply chain performance statically. However, supply chain management strategy was a weak predictor of supply chain management performance.

Deshmukh (2012) studied that the relationship between sugarcane growers and millers at a specific management level, the mill supply area. The relationship will be analyzed through the management of cane flows from the growers' fields to the mill as a way to enhance the mill area profitability. In the proposed Supply Chain Methodology an attempt will be made to minimize harvesting to crushing time to obtain the maximum possible sugar recovery to enhance the profitability of an organization. For this purpose, cane growers and millers are needs to be treated as an inter-dependent enterprise and make an attempt to develop a systematic approach for analyzing the interfaces between the sugar industry and the farmers.

Odoom (2012) identified and described the way logistics and supply chain management practices could be used for costs savings. Every hotel in order to maintain and operate their properties, they need to emphasis on issues like information technology, procurement and distribution, logistics and inventory management. Finally, the paper explored the competitive benefits emerging from applying these concepts. Hence, hotel managers needed to focus their attention on changing customer demands and should be able to identify the costs benefits that could be derived from utilizing logistics and supply chain management costs saving strategies. The hotel industry to gain the competitive advantage should implement and execute supply chain management properly. Supply chain management should act as a guide to gain the competitive advantage.

Mangal (2013) Supply Chain Management is a systematic approach to improve the total productivity of the sugar industries by optimizing the timing, location and quantity of material flow from sugar cane to sugar at consumer's site using IT infrastructure and interacting with all the related intermediates and that is expected approach for implementing Total Quality Management philosophy to improve organization. Supply Chain Management and Total Quality Management are the ways to cost optimization one, all along the chain, while other related to total business, but both starting from customer order to the delivery of goods to him.

Reddy and Raju (2013) highlighted the present issues and challenges of Supply Chain Management in India. Supply chain management is a study which stresses on the best possible use of the resources and delivering the goods on time from supplier to the consumers. Companies become competitive and profitable with an efficient supply chain. In most of the cases, an organization achieving a good Supply Chain Management practices is definitely a challenging task because of changes and implementation of so many government policies, business policies, technology changes and the capabilities of the company, transportation, infrastructure, etc. In this study, the researchers have explored the importance and issues related to supply chain management of India. Finally, this study tried to suggest most of the Indian organizations to improve performance by applying and implementing the techniques which offer a strategic opportunity for the companies to increase its revenues. This can be achieved by integrating IT with supply chain management, logistics and on the other side, develop the infrastructure required to integrate all these activities. Organizations must realize that they must harness the power of IT to collaborate with their business alliances.

❑❑❑

Chapter 3

Supply Chain Management Agri-Business

Case of Basmati Rice Brand

3.1 Introduction

Supply chain management is helpful in reducing costs. It provides better medium for information sharing between partners. It is also helpful in improving customer's satisfactions as well as services. Supply chain management is also helpful in increasing cashflow and improves quality and gives higher profit margin. The main goal of this study is to identify cause and effect relationship between dependent and independent variables. The sampling structure and techniques adopted during the course of investigation have been described in this chapter.

3.2 Scope of the Study

The study has been confined to different groups and all possible local areas in Jammu city. The sample size was of 150 respondents, which included people from various age groups and different income levels. This study aims towards collecting information about supply chain management of selected basmati rice unit and thus, helping in understanding the effectiveness of its supply chain and help in taking new and better decisions which will, thus help them in boosting their business. This study basically highlights the ideas about what company should do at the condition of unbalance and haphzardness in their supply chain. By this study, we came to know how the organization retaining and building a good relationship between retailers, wholesalers, consumers through their supply chain management channel. The main purpose of this project was to study the supply chain system and supply chain problems/constraints faced by channel partners of "Selected Basmati Rice mill" so as to develop strategies in order to boost the supply chain system with the company to increase the market share by proper and smooth delivery system mechanism.

3.3 Objectives of the Study

1. To identify the supply chain aspect of selected basmati rice unit in Jammu region.
2. To identify the key constraint in supply chain management of basmati rice in Jammu region.

3.4 Limitations of the Study

1. The study relates only to the consumer of selected areas of Jammu city of J&K, therefore, the findings cannot be true representation of all consumers (total population).
2. The personal bias of respondent has also been a matter of concern.
3. Respondent's ignorance to certain questions also posed as barrier towards certain responses.
4. Since the researcher has used structured schedule, the respondent's responses were immediate and the time taken for deciding was less, which led to responses with not much thought. Therefore, time constraint has also been a limitation in this study.
5. Illiteracy of the respondents posed a barrier for making them understand the question. Therefore, the responses cannot be taken factual.

3.5 Methodology

3.5.1 Locale of Study

The research was conducted with the help of schedule based on the information collected from the respondents in various geographical area of Jammu city.

3.5.2 Type of Study

Although there are various methods of research studies mentioned in literature for conducting study in different disciplines. This study has been based on descriptive type of research. A research design provides the framework to be used as the guide in analyzing the data experiences with different research design will generally provide the capability to match a research problem with an appropriate design.

3.5.3 Sampling Technique

The present study has adopted Convenience sampling technique to conduct the survey among the respondents.

3.5.4 Sample Size

The total sample of the study was 150 respondents. Out of sample, Farmers consist of 50, Sarveshwar employees (34), Wholesalers (6), Retailers (60) of Sarveshwar Basmati brand in different areas of Jammu city. The data has been collected through filling up the schedule from various respondents' viz., farmers, Sarveshwar employees, wholesalers and retailers.

3.5.6 Data Collection

The data collection is the method to collect important information to keep on record for further use, to make important decisions about different issues and is of vital significance for others. The present study has adopted both the primary as well as the secondary data collection techniques to conduct the present study.

- **Secondary Data**: The data which has already been collected, complied and presented earlier by any agency was used for purpose of investigation. The data was collected through websites, journals related to brand and advertisement, newspapers, magazines, etc.
- **Primary Data**: it involved the first hand information collected through schedule by direct interviews.
  a) It contains responses from the Farmers, Wholesalers, Retailers and Consumers.
  b) Variable analyzed.
- Supply chain practices and information system.

3.5.7 Variables involved in the Schedule

For objective 1: For identification of supply chain of Basmati rice in Jammu region, following statements were measured and analyzed and the results were calculated by using flow chart technique:

1. Firm's information system capability
2. Customer information
3. Just in time/supply chain practices

For objective 2: For identification of key constraints in supply chain management of Basmati rice in Jammu region, following statements were measured and analyzed and the results were calculated by using regression, descriptive, statistics:

3.6 Methods of Analyzing Data

After collecting the data, the results had been analyzed by using descriptive statistics, percentage analysis and regression analysis technique. Descriptive statistics mainly mean, median, mode, standard deviation was used to minimize large number of variables into meaningful and correlated form.

3.6.1 Percentage Analysis

Percentage refers to a special kind of ratio. It is used to make comparison between two or more series of data. They can be used to compare the relative items, the distribution of two or more series of data, since the percentage reduces everything to a common base and there by allow meaningful comparisons to be made.

Percentage = (x/y) X (100/1)

Where x = number of respondents respond

y = total number of respondents

3.6.2 Arithmetic Mean

The arithmetic mean is calculated by adding a group of numbers and then, dividing by the count of those numbers,

Mean = " $(x_1, x_2 \ldots x_n)/n$

Where, $x_1, x_2 \ldots x_n$ are the observation

n = number of respondents

3.6.3 Median

Median which is the middle number of a group of numbers; that is, half the numbers have values that are greater than the median, and half the numbers have values that are less than the median.

$$\text{Median} = \frac{\sum (N+1)}{2} \text{th item}$$

3.6.4 Standard Deviation

The standard deviation is a measure of how widely values are dispersed from the average value (the mean).

$$\text{Standard Deviation } (\sigma) = \frac{\sqrt{\sum (\text{x-Mean})^2}}{(\text{n-1})}$$

Where, n is the sample size.

3.6.5 Regression Analysis

Regression takes a group of random variables X's (independent variables), thought to be predicting Y (dependent variable) and tries to find the mathematical relationship between them. Multiple regression analysis technique will be adopted in the present contribution to establish the relationship among the variables.

Model to be adopted to determine Supply Chain

Y_1=f(X1, X2, X3)

Y_1= a+β1(X1) +β2(X2) +β3(X3)

Where,

- $Y_1 = f(x_1, x_2, x_3)$

  Where Y_1 = Supply chain Practices & x_1, x_2, x_3 are questions of Supply Chain Practices.

  Where x_1 = Supply chain planning, x_2 = Just-in-time and x_3 = Delivery practices.

- $Y_2 = f(\mu_1, \mu_2, \mu_3, \mu_4)$

  Where Y_2 = Information Sharing System & μ_1, μ_2, μ_3, μ_4 = questions of Information system

  Where μ_1 = Information system, μ_2 = Customer information,

  μ_3 = Manufacturing Information and μ_4 = Information quality .

  Or

- $Y_1 = a + \beta_1 (SCP) + \beta_2 (JIT) + \beta_3 (DP)$
- $Y_2 = a + \beta_1 (IS) + \beta_2 (CI) + \beta_3 (MI) + \beta_4 (IQ)$

Note

- IS = *Information System*
- CI = *Customer Information*
- MI = *Manufacturing Information*
- IQ = *Information Quality*
- SCP = *Supply Chain Planning*
- JIT = Just in Time
- DP = Delivery Practices

3.7 Results

3.7.1 Product ranges of Sarveshwar Foods Limited

Table 1 Showed that product ranges of Sarveshwar Foods Limited. From this table, it is clear that the company has various product ranges categorized in basmati rice, non-basmati rice and premium basmati rice. In basmati rice, the company has basmati brown rice, basmati raw rice, basmati white parboiled rice and basmati golden parboiled rice. In non-basmati rice, the company has Indian long/medium/short grain white raw rice, Indian long/medium/short white parboiled, Indian long/medium/short golden parboiled rice, Indian long/ medium/short brown rice. In Premium Basmati Brands, the company has Sarveshwar Grand Basmati Rice, Sarveshwar Unique Basmati Rice, Sarveshwar Ultra XL Basmati Rice, and Sarveshwar Select Basmati Rice.

Table 1. Product ranges of Sarveshwar Foods Limited

Basmati Rice	Non- Basmati Rice	Premium Basmati
Basmati Brown Rice	Indian Long/ Medium/ Short White Raw Rice	Sarveshwar Grand Basmati Rice
Basmati Raw Rice/ Silky Raw Rice	Indian Long/ Medium/ Short White Parboiled	Sarveshwar Unique Basmati Rice
Basmati White Parboiled Rice	Indian Long/ Medium/ Short Golden Parboiled Rice	Sarveshwar Ultra XL Basmati Rice
Basmati Golden Parboiled Rice	Indian Long/ Medium/ Short Brown Rice	Sarveshwar Select Basmati Rice

Source: *Sarveshwar Foods Limited*

Table 2 showed various testing equipments that are being used by the company for checking the quality attributes. The table showed that for measuring the length and testing the moisture, Vernier caliper and Moisture meter is used in the Sarveshwar Foods Limited Laboratory. The table also depicts that for Whiteness testing, Paddy testing, Rice polishing and final cooking, Whiteness meter, Lab husker, Lab polisher and Rice cooker is used for the whole process.

Table 3 represented that the total number of plants and machineries of the company. The company has six (06) Intake Conveying, four (04) Silos, one (01) Pre Cleaner, two (02) Par Boiling Plant, three (03) Dryers, ten (10) Elevators, one (01) Boilers, one (01) DE Mineralization Plant, one (01) Paddy Cleaner, one (01) DE Stoner, two (02) DE Husker, one (01) Paddy Separator, two (02) Thickness Grader, four (04) Polisher, two (02) Silky, one (01) Colour Sorter, twenty six (26) Length Blader, eight (8) Magnets, twenty eight (28) Elevators, four (04) Conveyors, one (01) Consumer Packing, one (01) Bulk Packing, one (01) Wrapping and Strapping, three (03) Elevators, and one (01) Conveyors . Intake Conveying, Silos, Pre Cleaner, Par Boiling Plant, Dryers and Elevators are located in the Parboiling Point. Boilers and DE Mineralization Plant are located in the Boiler. Paddy Cleaner, DE Stoner, DE Husker, Paddy Separator, Thickness Grader, Polisher, Silky, Colour Sorter, Length Blader, Magnets, Elevators, Conveyors and Consumer Packing are located in the Milling Plant. Consumer Packing, Bulk Packing, Wrapping and Strapping, Elevators and Conveyors are located in the Packing Plant.

Table 2. List of Testing Equipments

Length measurement	Vernier calliper
Moisture testing	Moisture meter
Whiteness testing	Whiteness meter
Paddy testing	Lab huskers
Rice polishing	Lab polishers
Final cooking	Rice cookers

Source: Sarveshwar Foods Limited

Table 3. Plant and Machinery of Sarveshwar Foods Limited

S.No.	Name	Location	Quantity
1	Intake Conveying	Par-boiling Point	06
2	Silos		04
3	Pre Cleaner		01
4	Par Boiling Plant		02
5	Dryers		03
6	Elevators		10
7	Boilers	Boiler	01
8	DE Mineralization Plant		01
9	Paddy Cleaner	Milling Plant	01
10	DE Stoner		01
11	DE Husker		02
12	Paddy Separator		01
13	Thickness Grader		02
14	Polisher		04
15	Silky		02
16	Colour Sorter		01
17	Length Blader		26
18	Magnets		08
19	Elevators		28
20	Conveyors		04
21	Consumer Packing	Packing Plant	01
22	Bulk Packing		01
23	Wrapping and Strapping		01
24	Elevators		03
25	Conveyors		01

Source: Sarveshwar Foods Limited

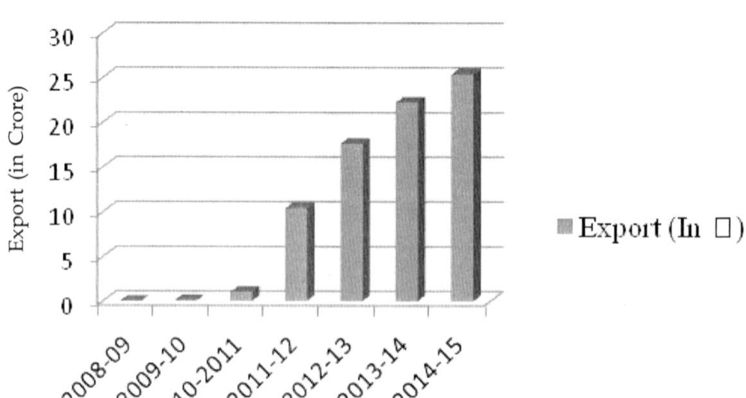

Figure 1: Export of Basmati Rice

Table 4 and Figure 2 represented the year wise production of Sarveshwar Foods Limited. It is clear from this table that the annual production level of the Basmati has increased over the years. The table showed that in the year 2009-10, 2010-11, 2011-12, 2012-13, 2013-14, 2014-15 and 2015-16, the total production of the company was calculated as 99.30 thousands Metric tonnes, 101.30 thousands Metric tonnes, 98.56 thousands Metric tonnes, 101.35 thousands Metric tonnes, 94.70 thousand Metric tonnes, 110.40 thousands Metric tonnes and 122.50 thousands Metric tonnes respectively.

Table 4. Year wise Production of Selected Agri-Business Basmati Mill

Year	Production (000 MT)
2009-10	99.30
2010-11	101.30
2011-12	98.56
2012-13	101.35
2013-14	94.70
2014-15	110.40
2015-16	122.50

Source: Sarveshwar Foods Limited

3.7.2 Supply Chain of selected Basmati Rice Mill

The Flowchart 1 represented the Supply Chain of selected Basmati Rice mill. The flowchart showed that the company has a well based supply chain channel, through which the supply and distribution process takes place. The flowchart 1 revealed that the process of supply starts from farmers or procurement agencies, after that the grains or paddy is transferred from the procurement agencies/farmers in two different processes, viz. supply to the rural assemblers and miller/processor as well. The paddy/grains collected through rural assemblers are also transferred to the miller/processor. After processing is done, the miller supplies the finished product to the wholesalers, retailers, directly to the consumers and also supplies (export) out of the region. The wholesalers then supply and distribute these products (Basmati) to the retailers and the consumers directly. The retailers at last sell these products to the ultimate consumers for consumption purposes.

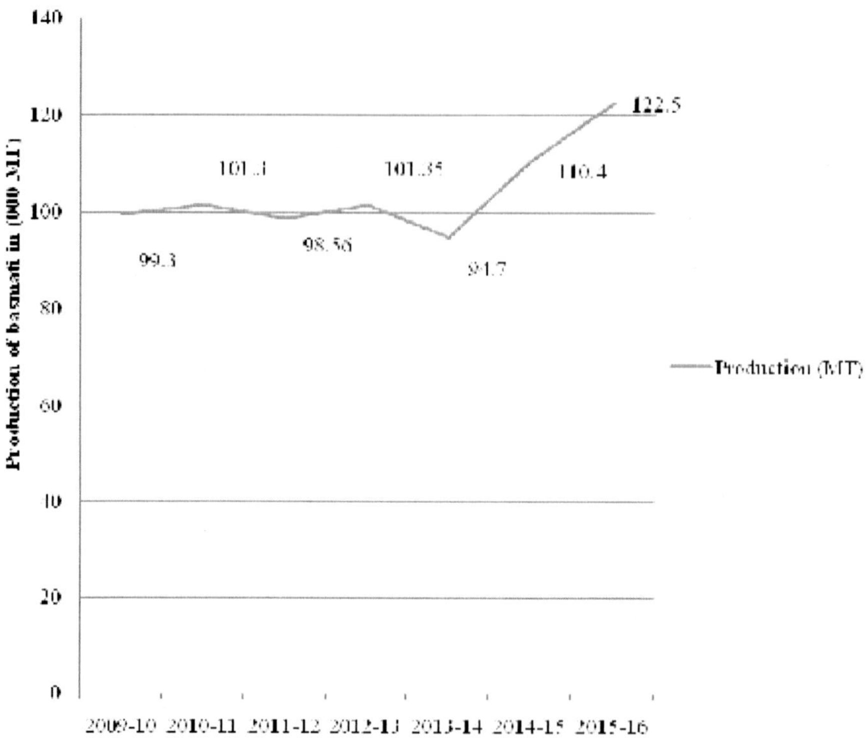

Figure 2: Year wise Production of selected Basmati Rice Mill

3.7.3 Activities of Supply Chain actors

Table 4 represented the main activities performed by the Chain actors. The table showed that the main activity of the farmers is Harvesting. Collectors performs the activity of Paddy collection. Grading, processing, packaging and labelling activities are performed by Millers, Wholesalers are engaged in the activity of Bulk stocking, and the distribution process is performed by retailers and the ultimate consumption by the Consumers.

3.7.4 Showing Supply Chain (Elements) Relationship

Flowchart 3 represented the supply chain relationship between various factors. The flowchart revealed that the elements of information sharing system with the calculated R value viz. information system (0.65), customer information (0.61), manufacturing information (0.87), showed a positive relationship with the information system at large extent, which means there is inter-dependency between the information sharing system with the supply chain process and similarly, supply chain practices (dependent variable) showed positive relationship with the inter-dependent variables viz. supply chain planning (0.28), just in time (0.195), delivery practices (0.389) respectively.

Thus, the flowchart model shows that there is a inter-dependency between dependent and inter-dependent variables, hence the supply chain system of the Sarveshwar Foods Limited has acquired a positive relationship with the information system and supply chain practices respectively.

3.7.5 Re-Designing Model of Supply Chain for selected Basmati Rice Mill

The Flowchart 4 represents the re-design model of Supply chain of selected Basmati Rice mill. This re-design model is being framed while going through the result of the study. The flowchart reveals that the company should adopt the following steps taken in this flowchart for better performance of supply chain of basmati rice. This flowchart again regards supply chain that starts from supply from farmers in two ways i.e. supply of paddy from large landholding and supply of paddy through small landholding. Farmers with large landholding directly supply their produced to the rice processing unit, where the process of paddy cleaning, de-husking, polishing, packing and storage is done. Whereas, farmers with small landholding can supply their produced to the procurement centers of Sarveshwar Foods Limited, which are opened in the villages in order to collect the paddy from the small landholding farmers in small quantities and after collecting the paddy from the small landholding farmers, the procurement centers transferred these paddy to the rice processing unit for processing. After processing is done, the finished products is directly supplied to the retailers by eliminating the wholesaler's role from the chain, the finished products also supplied for export purposes, the products supplied to the retailers are ultimately sold to the consumers for consumption purposes.

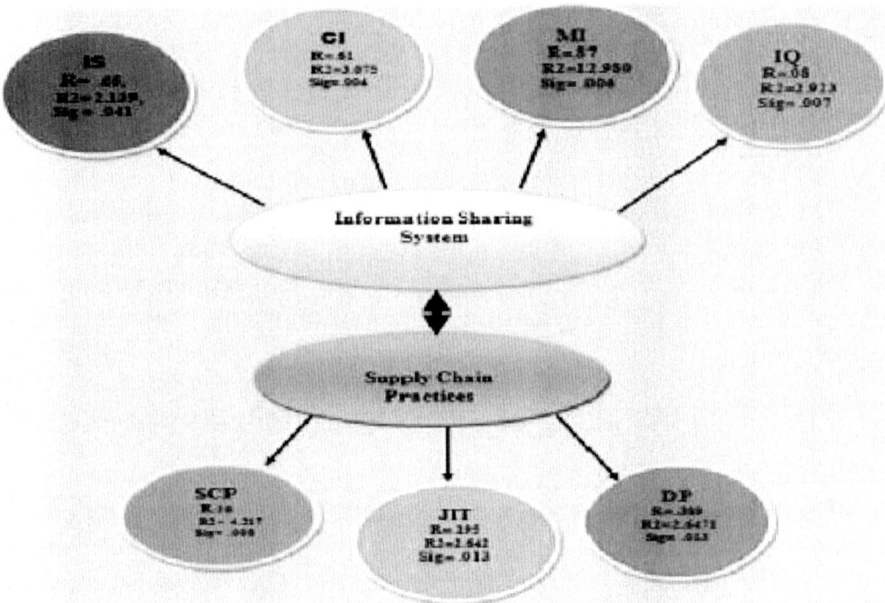

Flowchart 3.1 Supply Chain (Elements) Relationship

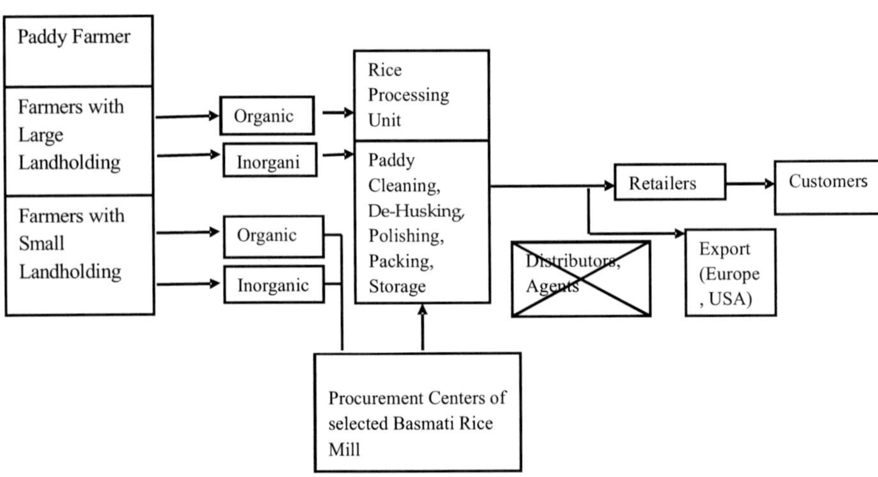

Flowchart 3.2 *Re-designing Model of Supply Chain for selected Basmati Rice Unit*

Table 5. Activities of Supply Chain actors

Chain Actors	Activities
Farmers	Harvesting
Collectors	Paddy Collection
Miller	Grading, Processing, Packaging and Labelling
Wholesalers	Bulk Stock
Retailers	Distribution
Consumer	Consumption

Source: Survey

3.7.6 Problems and Constraints Faced by Farmers

Table 5.10 and figure 5.6 represented the main problems and constraints faced by farmers in supply chain of basmati rice. It is clear from this table that logistic (26 per cent) is the main problems faced by the farmers followed by problem of payment (22 per cent), problem of lack of institutional support (20 per cent), storage problem (18 per cent), and problem of price setting (14 per cent) respectively.

3.7.7 Constraints faced by Wholesalers

Table 5.11 represented the main constraints faced by the wholesalers. From this table, it is clear that the main constraints faced by the wholesalers is in getting license (2.50) followed by lack of information and competition (2.00), then high tax rate (1.75) and then lack of capital (1.50).

3.7.8 Constraints faced by Retailers

Table 5.12 represented the main constraints faced by the retailers. It is clear from this table that the main constraints faced by the retailer is shortage of supply of rice (3.23), followed by storage (2.87), adulteration (2.77), capital shortage (2.73), quality (2.63) and taxation and other fees (2.33).

3.7.9 Constraints faced by Milling Unit

Table 5.13 represented the main constraints faced by Milling Unit. It is clear from this table that lack of control on farmers (3.47) is the main constraints faced by the company, followed by lack of improved rice storage facilities and lack of technology & Human Resource (3.29 both), governmental support (2.32).

Table 6. Problems and Constraints Faced by Farmers

Constraints	Percentage	Frequency
Logistics	26	13
Storage	18	9
Payment	22	11
Lack of institutional support	20	10
Problem of price setting	14	7
Total	**100**	**50**

Source: Survey

Table 7. Constraints Faced by Wholesalers

	Lack of Capital	Lack of Information and Competition	High tax rate	License Procedure
Mean	1.50	2.00	1.75	2.50
Standard Error	0.29	0.41	0.25	0.65
Median	1.50	2.00	2.00	2.50

Source: Survey

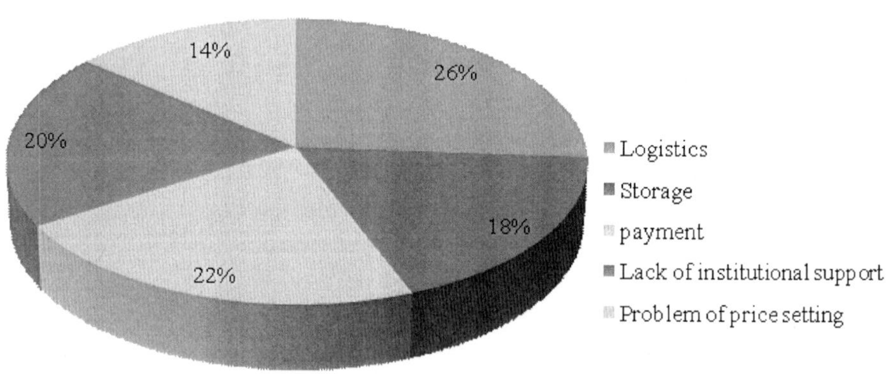

Figure 3. Problems and Constraints Faced by Farmers

Table 8. Constraints Faced by Retailers

	Taxation and other Fees	Shortage of Supply of Rice	Storage	Quality	Adulteration	Information Flow	Capital Shortage
Mean	2.33	3.23	2.87	2.63	2.77	2.42	2.73
Standard Error	0.15	0.14	0.14	0.13	0.12	0.13	0.13
Median	2.00	4.00	3.00	3.00	3.00	2.00	3.00
Mode	2.00	4.00	2.00	2.00	2.00	2.00	2.00
Standard Deviation	1.17	1.09	1.08	1.02	0.96	1.00	1.02
Sum	140.00	194.00	172.00	158.00	166.00	145.00	164.00
Count	60.00	60.00	60.00	60.00	60.00	60.00	60.00

Source: Survey

Table 9. Constraints Faced by Milling Unit

	Lack of Improved rice Storage facilities	Lack of Technology and HR	Lack of Control on Farmers	Governmental Support
Mean	3.29	3.29	3.47	2.32
Standard Error	0.15	0.13	0.11	0.11
Median	3.00	3.00	4.00	2.00
Mode	4.00	4.00	4.00	2.00
Standard Deviation	0.87	0.76	0.61	0.64
Sum	112.00	112.00	118.00	79.00
Count	34.00	34.00	34.00	34.00

Source: Survey

❑❑❑

Chapter **4**

Summary and Conclusion

The research entitled *"Supply Chain Management in Agri-Business: Case of Basmati Rice Brand"*was carried out in Jammu region, 2016. A total of 150 respondents were selected from the sample area. The information was collected from in the sample area. The data collected was subject to analysis for examining the objectives of the investigation. The findings of the present investigation have been briefly summarized in this chapter.

The company has a well based resource structure, operational and tactical management teams. The result concluded that company has business dealings with three main types of product ranges viz. Basmati rice, Non-Basmati rice and Premium Basmati rice. The current on roll employees of the Sarveshwar Foods Limited is 91 in number in the year 2015-16. The original investment in plant and machinery in the year 2014-15 was calculated as 3.64 crore. The company has a role in export and is exporting its products to the European countries like USA. The company also has a well updated plants and machineries, various Testing equipments for checking the quality attributes. The result calculated that Annual production level of the Sarveshwar Basmati has increased over the years, as was calculated as 122500 Metric tonnes in 2015-16.

The company has a well based supply chain channel through which the supply and distribution process takes place and the chain comprises of miller, farmers, wholesalers, retailers, and end consumers and is also known for chain actors of selected rice mill. In the re-designing model of supply chain of selected rice mill, the company should adopt the re-designing model for better performance of supply chain of basmati rice and also for the strengthening and boosting of supply chain of selected rice mill.

The study stated that the main constraints faced by the farmers in the sample area resulted to be logistics, means that the company should provide the logistics facilities in order to overcome logistics problem. Similarly, the main problem faced by the wholesaler, retailers and milling unit were found to be the problem of getting license, problem of shortage of supply of rice and lack of control on farmers respectively.

Conclusion

The study concluded that the production of the basmati rice is calculated as 8773.78 thousands tonnes in 2014-15, as was improved and greater from the previous year 2013-14, which was 6616.13 thousand tonnes. The present study revealed that the selected Basmati Rice mill unit is having a well established "SCM" viz. having a sound managing body, human resource and processing plan as well. In this present study, the Supply Chain Management actors identified are farmers, miller, wholesalers, assemblers, retailers and consumers. In terms of Supply Chain, the company has a sound channel partners. Due to which the company has achieved new heights in the market and has acquired 75 per cent of market share in the sample area.The study also depicts that among the various factors logistics, license procedure, lack of control on farmers and unfulfillment of demand in the market are identified as the key constraints in the Supply Chain Management.

❑❑❑

Bibliography

Agarwal, A. and Shankar 2002. "Analyzing alternatives for improvement in supply chain performance". *Work Study*, **51**(1): 32–37.

Armstrong M, Brown D 2008. Reward strategies and trends in the United Kingdom: the land of diverse and pragmatic dreams Compens. Ben. Rev., **37**(4): 41-53.

Amaresha M. ,Dinakar G., Nagesha B.N., 2010. "Role of Logistics and Supply Chain Management in Organized Retail, **2**(3): 232-239

Baba, V. V. and Jamal, M. 1991. "Routinisation of job context and job content as related to employees quality of working life: A study of psychiatric nurses". *Journal of organizational behaviour*, **12**: 379-386.

Bhardwaj, S. B. L. 1982. "QWL Perspectives, Dreams and Realities", Paper Presented at National Symposium on QWL, Mimeographed, Hyderabad.

Buchanan, D.A., and Boddy, D., 1983. "Advanced Technology and the Quality of Working Life: The Effects of Computerized Controls on Biscuit-Making Operators". *Journal of Occupational Psychology*, **56**: 109-119.

Bino, Thomas, 1995. "Impact of Quality Circles-A Study with Reference to the Manufacturing Industries in Kerala", Ph.D. Thesis, Cochin University of Science and Technology, Cochin.

Bowersox, Donald J. and David C. Closs 1996. Logistical Management: *The Integrated Supply Chain Process*, McGraw-Hill Series in Marketing, New York: The McGraw-Hill Companies.

Cohen, M. A. and H. L. Lee, 1985. Manufacturing Strategy Concepts and Methods, in Kleindorfer, P. R. Ed., *The Management of Productivity and Technology in Manufacturing*, 153-188.

Cohen, M. A. and H. L. Lee, 1989. "Resource Deployment Analysis of Global Manufacturing and Distribution Networks", *Journal of Manufacturing and Operations Management*, 81-104.

Christopher, Martin L. 1992. Logistics and Supply Chain Management, London: Pitman Publishing.

Coase, R. H, 1998. The New Institutional Economics', *AEA Papers and Proceedings*, May, 72-74.

Chen, I. J. and Paulraj, A, 2004. "Towards A Theory of Supply Chain Management: The Constructs and Measurements", *Journal of Operations Management,* **22**(2): 119-50.

Chen, I. J. and Paulraj, A, 2004. "Understanding Supply Chain Management: Critical Research and a Theoretical Framework", *International Journal of Production Research,* **42**(1): 131-163.

Chin, K.S., Tummala, V.M.R., Leung, J.P.F. & Tang, X., 2004. A study on supply chain management practices: The Hong Kong manufacturing perspective. *International Journal of Physical Distribution & Logistics Management,* **34**(60): 505-524.

Chopra, S. and Meindl, P., 2007. Supply Chain Management: Strategy, Planning and Operation 2nd or 3rd., New Jersey: Pearson Prentice Hall.

Corsten, D. & Gruen T, 2007. Desperately seeking shelf availability: An examination of the extent, the causes, and the efforts to address retail out-of-stocks. *International Journal of Retail & Distribution Management,* **31**(11/12): 605-617.

Drucker, Peter F. 1998. "Management's New Paradigms," Forbes Magazine, 152-177.

Duclos, L. K.., Vokurka, R. J., and Lummus, R. R. 2003. "A Conceptual Model of Supply Chain Flexibility", Industrial Management & Data Systems, **103**(6): 446-456.

Duncan Gallie, 2003. "The Quality of Working Life: Is Scandinavia Different?" *Oxford Journal,* **19**: 61-79.

Dr Dharamvir Mangal, 2013. "Supply Chain Management-A Quality Improving Tool in Process Industries", *International Journal of Application or Innovation in Engineering & Management* (IJAIEM), **2**(1).

Ellram, L., Cooper, M., 1993. "Characteristic of Supply chain Management and the Implications for Purchasing and Logistics Strategy", *International Journal of Logistics Management.* **4**(20): 13-24.

Fawcett, S.E, 1992. "Strategic Logistics in Coordinated Global Manufacturing Success", *International Journal of Production Research,* **30**(5): 1081-2000.

Hofstede, G, 1980. Culture's Consequences, Sage, Beverly Hills, CA. 119-130.

Hofstede, Geert, 1980. Culture's Consequences: International Differences in Work Related Values. London: Sage.

Hatrenstien, A., and Huddleston, K.F., 1984. "Values: The Cornerstone of QWL", Human Relations, **38**(10): 65-66.

Harland, C.M., 1996. "Supply chain management relationships, chains and networks", *British Journal of Management*, **7**: 63-80.

Harland, C.M, 1996. Supply Chain Management, Purchasing and Supply Management, Logistics, Vertical Integration, Materials Management and Supply Chain Dynamics, in Slack, N (ed.) Blackwell Encyclopedic Dictionary of Operations Management. Blackwell, 1996.

Heskett, J., W. E. Sasser Jr., and L. Schlesinger, 1997. "The Service Profit Chain", N.Y.: Free Press

Hines, T. 2004. Supply chain strategies: Customer driven and customer focused. Oxford: Elsevier.'

Joseph, Zakhariya, P.J., 1999. "Employees Perception of their Quality of Work Life in Apollo Tyres Ltd", Kalmassery, MBA Dissertation, Cochin University of Science and Technology, Cochin.

Kalse, R.V., Kalyankar, S.P. and Patil, H.N. 1996. Agro processing industries: Existing potential and capacity utilization in Maharashtra. Indian J.agric.Mktg., **10**: 54-55.

Kaplinsky R, 2000. "Spreading the gains from globalization: What can be learned from value chain analysis?" *Journal of Development Studies*, **37**(2): 117-146.

Koprulu, A. and Albayrakoglu, M. M. 2007. Supply chain management in the textile industry: a supplier selection model with the analytical hierarchy process, Viña Del Mar, Chile, **3**(6).

Lawler, E., 1982. "Strategies for Improving the Quality of Work Life", *Journal of Applied Social Psychology*, **37**(5): 486-493.

Lee, H.L. & Billington, C., 1993. Material Management in Decentralized Supply Chains, Operations Research, **41**: 835–847.

Lummus, R. R., Duclos, L. K., and Vokurka, R. J. 2003. "Supply Chain Flexibility: Building a New Model", *Global Journal of Flexible Systems Management*, **4**(4): 1-13.

Li, S., Rao, S.S., Raghunathan, T.S. & TaguNathan, B., 2005. "Development and Validation of a Measurement Instrument for Studying Supply Chain Practices," *Journal of Operations Management*, **23**: 618-41.

Li, S., Raghunathan, B., Raghunathan, T.S. & Subba-Rao, S., 2006. The impact of supply chain management practices on competitive advantage and organizational performance. Omega, **34**(1): 107-124.

Lambert, D.M. 2008. "Supply Chain Management: Implementation Issues and Research Opportunities." *International Journal of Logistics Management*, **9**(2): 1-19.

Maccobby, Michael, 1984. "The Quality of Work Life in Bell System" Monthly Labour Review, **10**(3): 31-32.

Mirvis, P.H. and Lawler, E.E., 1984. "Accounting for the Quality of Work Life", *Journal of Occupational Behaviour*, **5**(3): 197-212.

Mathur, P. and Paranjape, S., 1989. "Identification of training areas for special educators as related to job satisfaction factors of QWL perceptions. Paper presented on 25th Annual Conference of Indian Academy of Applied Psychology.

Mentzer, J.T., 2001. Defining Supply Chain Management, *Journal of Business Logistics*, **22**(2): 1-25.

Mentzer, J. T., DeWitt, W., Keebler, J. S., Min, S., Nix, N. W., Smith, C. D., & Zacharia, Z.G. 2001, "Defining Supply Chain Management." *Journal of Business Logistics*, **22**(2): 1-25.

Madhavedi, S. & Acharyulu, G.V.R.K., 2010. Supply chain management practices in traditional fresh produce distribution system in India: an empirical study. *International Journal of Strategic Management*, **10**(2).

Nilakant, V. & Tandon, R., 1982. "An Alternative Approach for Improving QWL in India", Occasional Paper No.1, PECCE, New Delhi.

Nitish.R. De, 1984. "Interlinkage between Quality of Work Life and Quality of Life", Productivity, **22**(4): 89-91.

Narayan, Asit and Jha, Amarnath, 1992. "Quality of Work Life- Concepts and Process", *the Indian Journal of Commerce*, **12**(2).

Odoom, Clement K., 2012. "Logistics and Supply Chain Management in the Hotel Industry: Impact on Hotel Performance In Service Delivery", UNLV Theses/Dissertations/Professional Papers/ Capstones. Paper 1339.

Porter, M. E. 1985. *The Competitive Advantage: Creating and Sustaining Superior Performance*. NY: Free Press.

Powell, W, 1990. Neither Market nor Hierarchy: Network Forms of Organization. *Research in Organizational Behaviour*. **12**: 295-336

Power, D., 2005. Supply chain management integration and implementation: a literature review. Supply Chain Management: *An International Journal*, **10**(4): 252-263.

Petrovic-Lazarevic, S., Sohal, A., and Baihaqi, I., 2007. Supply Chain Management Practices and Supply Chain Performance in the Australian Manufacturing Industry. Australia, Monash University.

Plazibat, I. & Brajevic, S., 2009. Supply Chain Management in Retail Industry. *Business Logistics in Modern Management*, **9**: 133-140.

Quayle, M., 2003. "A study of supply chain management practice in UK industrial SMEs", Supply Chain Management: *An International Journal*, **8**(1): 79-86.

R.S.Deshmukh, N.N.Bhostekar, U.V.Aswalekar , V.B.Sawant, 2012. "Inbound Supply Chain Methodology of Indian Sugar Industry" *International Journal of Engineering Research and Applications* (IJERA), **3**(2).

Ramana Reddy, M.V and Raju, N.VG.S., 2013. "Issues and Challenges of Supply Chain Management in India", *International Journal of Mechanical and Production Engineering*, **2**(1).

Christopher, Martin L. 1992. Logistics and Supply Chain Management, London: Pitman Publishing.

Sekaran, U, 1981. Perceived Quality of Working Life in Banks in Major Cities in India, Prajanan, **14**(3): 273-284.

Sinha, S.K., and Sahaya, R., 1981. "Management of Co-operative Enterprises (Principles and Practices)", National Council for Co-operative Training (National Co-operative Union of India), New Delhi. **12**(2): 5-8.

Singh, H., 1983. "QWL and Quest from Meaningful Work," Paper Presented at National Symposium on QWL, Hyderabad.

Singh, J.P., 1983. "QWL Experiments in India: Trials and Triumphs," Abhigyan, **2**(3): 23-27.

Scott, C. and Westbrook, R. 1991. "New Strategic Tools for Supply Chain Management", *International Journal of Physical Distribution and Logistics*, **21**(1): 23-33.

Srivastava, A. K., and Banerjee, R., 1996. "Organizational Effectiveness Scale", In D. M. Pestonjee (Ed.) Third handbook of psychological and social instruments, New Delhi: Concept Publishing House.

Stern, L.W., El-Ansary, A.I., Coughlan, A.T. & Anderson, E. 2001. Marketing Channels. 6th ed. Prentice Hall. Upper Saddle River. New Jersey.

Simatupang, T. M. and Sridharan, R. 2002. "The Collaborative Supply Chain", *International Journal of Logistics Management,* **13**(1): 15-30.

Sahay, B. S. & Mohan, R., 2003. Supply chain management practices in Indian industry. *International Journal of Physical Distribution & Logistics Management,* **33**(7): 582-606.

Shergill, H. S., 2007. "Sustainability of Wheat–Rice Production in Punjab: A Re-examination", *Economic and Political Weekly,* **13**(52): 81-85.

Singh, R. et al., 2010. Relating Organized Retail Supply Chain Management Practices, Competitive Advantage and Organizational Performance. Vision: *The Journal of Business Perspective,* **14**(3): 173–190.

Sukati, I. et al., 2011. The Study of Supply Chain Management Strategy and Practices on Supply Chain Performance. Procedia - Social and Behavioural Sciences, **282**(40): 225–233.

Thackray, R.I., 1981. "The Stress of Boredom and Monotony, A Consideration of the Evidence", Psychosomatic Medicine, **43**(2): 165-176.

Tan, K. C., Lyman, S. B., & Wisner, J. D. 2002. Supply chain management: A strategic perspective. *International Journal of Operations and Production Management* **22**(6): 614–631.

Thatte, A. A. 2007. "Competitive Advantage of a Firm through Supply Chain Responsiveness and SCM Practices" Published, PhD Dissertation. The University of Toledo.

Venkatraman,N.,&Ramanujam,V, 1986. Measurement of Business Performance in Strategy Research: a Comparison Approaches. *Academy of Management Review,* **11**: 801-814.

Whipple, J. M.; Frankel, R., 2000, Strategic Alliance Success Factors, in: *Journal of Supply Chain Management,* **36**(2): 21-28.

Wilson, O., 2003. "Work Ethics of the Industrial Workforce –A Study with Reference to selected Public and Private Sector Enterprises in Kerala", Ph.D. Thesis, University of Kerala, Thiruvananthapuram.

Wang G., Huang, S.H. & Dismukes J.P. 2005. Manufacturing supply chain and evaluation. *The International Journal of Advanced Manufacturing Technology,* **25**: 93-100.

Wong, C.Y., Boon-itt, S., & Wong, C.W.Y., 2005. "The Contingency Effects of Environmental Uncertainty on the Relationship between Supply Chain Integration and Operational Performance", *Journal of Operations Management,* **29**(6): 604–15.

❑❑❑